Ceramic Book & Literature Service

119 Brentwood Street

Marietta, Ohio 45750, U.S.A.

Telephone: 614 - 374-9458

HIGHLY DISPERSED METALS

ELLIS HORWOOD BOOKS IN INORGANIC CHEMISTRY

METAL IONS IN SOLUTION
J. BURGESS, Department of Chemistry, University of Leicester

CHEMISTRY OF ORGANO-ZIRCONIUM AND HAFNIUM COMPOUNDS
D. J. CARDIN, University of Dublin, M. F. LAPPERT, University of Sussex, and C. L. RASTON, University of Western Australia

BIO-INORGANIC CHEMISTRY
R. W. HAY, University of Stirling

METAL AND METALLOID AMIDES
M. F. LAPPERT, FRS, University of Sussex, A. R. SANGER, Alberta Research Council, R. C. SRIVASTAVA, University of Lucknow, India, and P. P. POWERS, Stanford University, California

SOLUTION EQUILIBRIA
F. R. HARTLEY, Royal Military College of Science, Shrivenham, C. BURGESS, Glaxo UK Limited, and R. M. ALCOCK, Severn-Trent Water Authority

INORGANIC SOLUTION CHEMISTRY
J. BURGESS, Department of Chemistry, University of Leicester (In preparation)

HIGHLY DISPERSED METALS

WŁADYSŁAW ROMANOWSKI
Head of Department of Catalysis
Institute of Low Temperature and Structure Research
Polish Academy of Sciences

Translation Editor:
Dr. Philip Gaskell
Churchill College, Cambridge
and
The Cavendish Laboratories,
University of Cambridge

ELLIS HORWOOD LIMITED
Publishers • Chichester
Halsted Press: a division of
JOHN WILEY & SONS
New York · Chichester · Brisbane · Toronto
PWN — Polish Scientific Publishers
Warsaw

English edition first published in 1987 in coedition between
ELLIS HORWOOD LIMITED
Market Cross House, Cooper Street, Chichester, West Sussex, PO19 1EB; England
and
PWN-POLISH SCIENTIFIC PUBLISHERS
Warsaw, Poland
The Horwood publisher's colophon is reproduced from James Gillison's drawing of the ancient Market Cross, Chichester
Revised translation from the Polish original
Metale w stanie wysokiej dyspersji, published in 1979 by Państwowe Wydawnictwo Naukowe, Warszawa—Wrocław

Distributors:

Australia, New Zealand, South-east Asia:
JACARANDA WILEY LTD
G.P.O. Box 859, Brisbane, Queensland 4001, Australia

Canada:
JOHN WILEY & SONS CANADA LIMITED
22 Worcester Road, Rexdale, Ontario, Canada

Europe, Africa:
JOHN WILEY & SONS LIMITED
Baffins Lane, Chichester, West Sussex, England

Albania, Bulgaria, Cuba, Czechoslovakia, German Democratic Republic, Hungary, Korean People's Democratic Republic, Mongolia, People's Republic of China, Poland, Rumania, the U. S. S. R., Vietnam, and Yugoslavia:
ARS POLONA-Foreign Trade Enterprise
Krakowskie Przedmieście 7, 00-068 Warszawa, Poland

North and South America and the rest of the world:
Halsted Press: a division of
JOHN WILEY & SONS
605 Third Avenue, New York, N. Y. 10158, U. S. A.

British Library Cataloguing in Publication Data
Romanowski, Władysław
Highly dispersed metals. — (Ellis Horwood series in inorganic chemistry)
1. Metals — Dispersion
1. Title II. Gaskell, Philip
669.9 TN673

Library of Congress Card No. 85-890
ISBN 0-85312-560-0 (Ellis Horwood Limited)
ISBN 0-470-20185-1 (Halsted Press)

Printed in Poland

Preface

This monograph is a slightly modified and updated version of the original Polish work *Metale w stanie wysokiej dyspersji* published in 1979 by Polish Scientific Publishers, Warsaw. It is intended to present a survey of the problems involved in the preparation, properties and uses of highly dispersed metals, which have had important industrial applications for a number of years and, more recently have aroused a vivid theoretical interest.

The references in the book should help the reader to continue his/her studies to the desired breadth and depth. Among the papers, cited a proportion (probably too high), refer to older papers and to the publications of Polish and Russian authors. This is deliberate. These publications, sometimes reporting pioneering and valuable work are less familiar to the Western scientific world, they demonstrate the fact that scientific interest in highly dispersed matter dates back to the nineteenth century.

In the studying the content of this book, nothing more than a rudimentary knowledge of chemistry and solid state physics is necessary. The "chemical slant" in the approach notwithstanding, it is hoped that this book will be helpful to all newcomers to the field, regardless of their previous experience and interest be it physical, or chemical.

Wrocław, May 1983 *W. Romanowski*

Table of Contents

Preface 5

Chapter 1 Occurrence and Application of Highly Dispersed Metals. Dispersion. Mean Dimensions and Distribution Functions

1.1 Introduction 11
1.2 Quantitative Description of Dispersion 13
1.2.1. Characterization of the dimensions of a single particle 13
1.2.2. Some definitions 16
1.2.3. The distribution of particle sizes 17
1.2.3.1 General properties of distribution functions 18
1.2.3.2 Some frequently-used distribution functions 20
1.2.3.3 Empirical distribution functions. 22
1.2.3.4 The log-normal distribution function 24
References 29

Chapter 2 Preparation of Highly Dispersed Metals

2.1 Elemental Metals 30
2.1.1. Preparation of higly dispersed metals by the decomposition of solids 31
2.1.1.1 Reduction of metal compounds by gaseous reducing agents 31
2.1.1.2 Reduction by solid reagents 34
2.1.1.3 Skeletal (Raney) metals 35
2.1.2 Reduction of metal ions in solution 36
2.1.2.1 Gold, silver and the platinum group metals 36
2.1.2.2 Non-noble metals 37

2.1.3 Preparation of dispersed metals from their gaseous compounds 38
2.1.4 Preparation of highly dispersed alloys 39
2.2 Physical Methods for the Preparation of Highly Dispersed Metals 40
2.2.1 Evaporation in an electric arc. 41
2.2.2 Evaporation of metals in an inert gas atmosphere . . . 41
2.2.3 Evaporation of metals in vacuum 45
2.2.4 Cathodic sputtering of metals 47
2.2.5 Other physical methods. 48
2.2.5.1 Formation of highly dispersed metal in solids by irradiation processes. 48
2.2.5.2 Low-temperature co-condensation of metals with non-metallic matrices (matrix isolation) 49
2.2.5.3 Implantation of metal into oxide substrates 51
2.2.5.4 Mechanical methods 51
2.3 Chemical Methods for the Preparation of Very Small Metallic Particles Deposited within and on the Surface of Solids 52
2.3.1 Iron, cobalt and nickel on carriers 54
2.3.2 Supported catalysts containing metals of the platinum group 55
2.3.3 Metal alloys deposited on carriers 56
References 57

Chapter 3 Physical Properties of Metals which are Sensitive to Dispersion. Application to the Determination of the Degree of Dispersion and the Structure of Small Metallic Particles

3.1 Electrical Properties of Dispersed Metals 61
3.1.1 Changes in the electrical conductivity during pressing and sintering of metal powder 62
3.1.2 The electric conductivity of metallic island films . . 63
3.1.2.1 Quantum mechanical tunnelling and electronic conduction in thin discontinuous metallic films 64
3.2 Magnetic Properties of Small Metal Particles 67
3.2.1 Ferromagnetic metals 67
3.2.1.1 Saturation magnetization and magnetic moment of small magnetic particles 67
3.2.1.2 Magnetization of systems containing small ferromagnetic particles 68
3.2.2 Paramagnetic metals 72

3.3 Interaction of Small Particles with Electromagnetic Radiation . . . 73
3.3.1 Scattering and absorption of light in a medium containing small particles . . . 73
3.3.1.1 Rayleigh scattering . . . 74
3.3.2 The broadening of X-ray diffraction lines . . . 77
3.3.3 Small angle X-ray scattering (SAXS) method . . . 80
3.3.3.1 Determination of the particle size distribution from small angle scattering . . . 82
3.3.4 Extended X-ray absorption fine structure (EXAFS) 83

References . . . 85

Chapter 4 Geometry and Electronic Structure of Small Metallic Particles

4.1 "Lattice" Structures of Small Atomic Aggregates . . . 89
4.2 The Stability of Small Atomic Aggregates . . . 91
4.2.1 Potential energy minimum . . . 91
4.2.2 Thermodynamic criteria for cluster stability . . . 93
4.3 Experimental Proof of the Existence of Metal Particles with and without Lattice Symmetries . . . 97
4.4 Electronic Structure of the Smallest Metal Particles . . . 99
4.4.1 Application of the methods of quantum chemistry to calculations of the electronic structure of metallic clusters . . . 100

References . . . 104
Appendix to Chapter 4 A Survey of Selected Quantum Chemistry Methods (by H. Chojnacki) . . . 106
References to the Appendix . . . 117

Chapter 5 Interaction of Small Metallic Particles with a Non-metallic Substrate and its Importance for Composite Catalysts and Thin Films

5.1 Determination of the Magnitude of the Interaction between Metal and Substrate Based on Measurements of the Contact Angle . . . 120
5.2 Experimental Determination of the Adhesion of Thin Metal Layers . . . 121

5.3 The Influence of Reactive Gases on the Adhesion . . . 123
5.4 Strong Interactions of Metals with the Substrate 124
5.4.1 Strong interactions in an atmosphere of a reactive gas 124
5.5 Electrical Phenomena at the Metal-Substrate Interface . . 126
5.6 The Interaction between a Metal and a Substrate and Its Effect on the Growth of Particles 127
5.6.1 Migration and coalescence of particles 128
5.6.2 Ostwald ripening 133
5.7 Interaction between the Metal and Substrate during the Nucleation of Metallic Particles 138
5.7.1 Behaviour of atoms on the substrate 138
5.7.2 Formation of nuclei 140

References 144

Chapter 6 Interaction between Gases and Highly Dispersed Metals. Adsorption and Catalysis

6.1 Highly Dispersed Metals as Adsorbents 146
6.1.1 Experimental techniques for investigating adsorption on metals 148
6.1.1.1 Low energy electron diffraction (LEED) 149
6.1.1.2 Investigation of the chemical composition of the surface by means of the Auger electron spectroscopy (AES) . . 151
6.1.1.3 Photoelectron spectroscopy (PES) and electron spectroscopy for chemical analysis (ESCA) 153
6.1.1.4 Field emission of electrons from metals 155
6.1.1.5 Electron- and thermally-stimulated desorption 158
6.1.1.6 Adsorption-induced changes in the function 161
6.1.2 Investigation of adsorption on highly dispersed metals 166
6.1.2.1 Adsorption equilibria 166
6.1.2.2 Infrared spectrometry of adsorbed molecules 170
6.1.2.3 The influence of adsorbed gases on the magnetic properties of highly dispersed metals 172
6.1.2.4 Adsorption-induced changes in the electrical conductivity of dispersed metals 174
6.1.2.5 Other methods 175
6.1.2.6 Adsorption on the smallest metallic particles 175
6.1.3 Application of the methods of quantum chemistry 176
6.1.3.1 The Ectended Huckel method 176
6.1.3.2 Other quantum chemical methods 180
6.2 Catalysis on Highly Dispersed Metals 181
6.2.1 Catalysis on solid surfaces 181

6.2.2 Catalytic activity of metals 183
6.2.2.1 Geometric factors in catalysis on metal surfaces . . . 183
6.2.2.2 Electron factors 185
6.2.2.3 Metal alloys as catalysts 186
6.2.2.4 Metallic molecular clusters in catalysis 188

References 191

Additional References 196

Index 204

1

Occurrence and Application of Highly Dispersed Metals. Dispersion. Mean Dimensions and Distribution functions

1.1 INTRODUCTION

The present book deals with "highly dispersed metals" which may be defined as metallic materials with definite chemical compositions (including therefore both elemental metals and their alloys), in the form of particles with very small dimensions. In general, the particles considered here are below 1 μm in size, and emphasis is placed on the smallest ones, with dimensions in the range of 1–100 nm.

Highly dispersed metals have played an important role in technology for some time. Their macroscopic chemical and physical properties, for instance, rates of reaction with other substances, colour, apparent density and rheological properties, differ significantly from bulk metals. The development of scientific research and technology in the last few decades has also prompted the application of highly dispersed metals, as well as highly dispersed solids in general.

In the present work, both common metals and precious metals will be considered. Among the few tens of elements in the periodic system known to be metals, the majority can be prepared in a highly dispersed state. However, the alkaline and alkaline earth metals, mercury, zinc, cadmium, tin, gallium, aluminium, indium and thallium only rarely occur in this state. These metals have few technical applications as highly dispersed metals since it is difficult to preserve them in this form at normal temperatures. They are more chemically active than other metals, and because of their comparatively low cohesion they tend to aggregate to form larger crystallites. Aluminium, in moderately dispersed form, is applied in protective coatings and decorations, where its passivity, that is the ability to form a thin, compact oxide layer, protects the interior of metal particle against environmental attack. Powders of barium, strontium and magnesium find some application, for instance in fireworks, where their rapid reaction with oxygen is utilized.

The highly dispersed solid state is thermodynamically unstable, thus highly dispersed metals are not found in nature, even in those deposits, where they are in the elemental state (copper, iron, noble metals). However, metals with high cohesive energy (high melting temperature) can be obtained and maintained in this state over long periods of time, since aggregation involves a high activation energy. At moderate temperatures, highly dispersed refractory metals are comparatively stable. Primarily, it is the transition metals and those of group IB which are most frequently obtained in a state of high dispersion and find various applications in this form.

The preparation of "colloidal" metals as sols and some of their properties and applications were known in the last century [1]. In addition, other forms of highly dispersed metals were obtained from their compounds by special methods, an example being the preparation of some transition metals by reduction of their compounds with metallic sodium. Very highly dispersed metals can also be obtained at moderate temperatures, for example, by reactions involving the carbonyls of iron, nickel and cobalt.

Colloidal metals may be prepared by chemical reduction of their compounds at ambient or slightly higher temperatures, or by formation of an electrical arc between metallic electrodes under water or another liquid. The first method may be applied to form colloidal noble metals, the second to produce refractory metals. It may be worth mentioning that it has been known for some time that colloidal metal particles have structures analogous to those of large crystallites and that they are not amorphous.

An important application of colloidally-dispersed metals is in photography. Here, development of the image formed by illumination of the photographic emulsion, involves the formation of very small particles of silver by the developer which acts, chemically, as a reducing agent. Gold and platinum emulsions find application in the decoration of glass and china. Thin metals films, obtained by condensation of a metal vapour in a vacuum consist of very small crystallites and now play an important part in electronics as getters, conductors and resistors and in optical devices such as mirrors, filters and semitransparent films.

Large quantities of highly dispersed metals, most often deposited on non-metallic carriers are now used in contact catalysis, without which contemporary chemical technology could not be imagined. Typical catalytic systems involve metals mainly from group VIII of the periodic table — the iron group metals, palladium, ruthenium and platinum, as well as copper and silver. The carriers (supports) on which the metallic particles are deposited are high area, porous oxidic materials which prevent aggregation of the dispersed metal into large crystallites. These supported metal catalysts are mostly obtained by reduction with hydrogen of the oxides or salts of the metal, previously deposited on the carriers.

To obtain certain refractory metals in bulk, it may be first necessary to resort to chemical preparation of very finely dispersed metallic powders from their compounds. These powders are subsequently pressed into various shapes and annealed (sintered) in a reducing or inert atmosphere or *in vacuo* at temperatures below the melting temperature (powder metallurgy). Examples are the manufacture of molybdenum, tungsten, osmium, iridium and platinum as well as nickel from nickel carbonyl. This provides an excellent method for the preparation of highly pure metals, as well as alloys and intermetallic compounds with special mechanical, electrical and thermal properties.

Some of the more recent techniques for the production of the less common metals such as nickel, cobalt and copper by hydrogen reduction or electrochemical deposition from aqueous ionic solutions (hydrometallurgy) can also yield metals with a high degree of dispersion.

1.2 QUANTITATIVE DESCRIPTION OF DISPERSION

1.2.1 Characterization of the dimensions of a single particle

The degree of dispersion of a solid can be described in terms of the linear dimensions of an individual particle. This is a quantity accessible to direct measurement with the aid of optical and electron microscopes but since the resolving power of an optical microscope is about 0.5 μm, it is of little use in investigations of the smallest particles. Electron microscopes, with a resolution of less than 0.5 nm, are clearly much more powerful. Determination of the dimensions of a particulate solid is, however, insufficient for characterization, since the ratio of the surface area of a sample to its mass depends on the shape of individual particles. It is thus difficult to give an accurate "geometrical" description of a system consisting of small particles, since dimensions and shapes can vary over wide limits. Very small particles may be characterized in practice by a quantity D, simply called *dispersion*, which is defined as the ratio of the number of atoms N_s, forming the surface, to the total number of atoms N in the particle

$$D = \frac{N_s}{N}. \tag{1.1}$$

Table 1.1. DIMENSIONS AND DISPERSION D OF SMALL POLYHEDRAL METALLIC PARTICLES COMPOSED OF VARIOUS NUMBERS OF ATOMS

Shape	N	Maximum dimension [nm]		N_s	D
Tetrahedron	35	Edge	1.25	34	0.97
	165		2.25	130	0.79
	1 140		4.5	580	0.51
	8 346		9.0	2 452	0.29
Cube	14		1.0	14	1.00
	172	Body	2.25	110	0.64
	1 099	diagonal	4.1	434	0.39
	7 813		7.8	1 730	0.22
Octahedron	19		1.05	18	0.95
	146	Diagonal	2.1	102	0.70
	1 156		4.2	486	0.42
	9 224		8.5	2 118	0.23
Cuboctahedron	13		0.7	12	0.92
	147	Height	1.55	92	0.63
	1 072		3.1	432	0.40
	8 000		6.2	1 734	0.22
Plate composed of 2 atom layers	arbitrary	Arbitrary		Equal to the total number of atoms	1.00

This quantity retains its meaning for systems composed of particles with different sizes and shapes. Numerically, D lies within the range from $D \simeq 0$ for "macroscopic" particles (with large dimensions) to $D = 1$ for very small particles, consisting of only a few atoms.

Dispersion, like the specific surface area (surface area per unit of mass) is however, an average quantity, giving only averaged data on the sizes of highly dispersed particles. Later, we consider in some detail more exact ways of describing systems containing particles with different sizes which are termed (only partially accurately) *polydisperse systems*. In this section, we show how the dispersion and specific area of *monodisperse systems* — that is, systems containing particles of uniform size — depend on the shape of the particles. We consider small (1–10 nm) metallic particles, in which the atoms are, typically, close-packed, filling approximately 74% of the space. The largest dimensions of regular bodies built from definite number of atoms of diameter 0.25 nm and dispersion D are given in Table 1.1. These data show that for metallic particles with a regular structure, D, decreases with increasing size and is dependent on the number

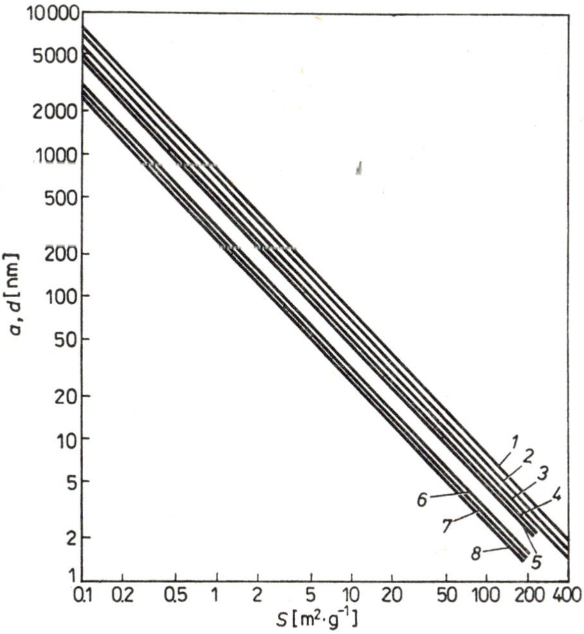

Fig. 1.1 — Dimensions of spherical or cubic particles plotted against specific surface area for some selected metals: *1* — Fe, *2* — Co, Ni, Cu, *3* — Mo, Ag, *4* — Pd, *5* — Rh, Ru, *6* — W, Au, *7* — Re, Pt, *8* — Ir.

of planes of polyhedra which defines the shape. Particles in the form of plates consisting of two (or less) atomic layers have all their atoms on the surface and $D = 1$. For particles with known dimensions one can easily determine their specific surface area assuming that they are spherical or cubic. In this case

$$S = \frac{6}{\varrho a}, \tag{1.2}$$

where S is specific surface area in $cm^2 \cdot g^{-1}$, ϱ is the density in $g \cdot cm^{-3}$ and a is the diameter of the sphere or the edge of the cube in centimetres.

Relation (1.2) is plotted in Fig. 1.1. for metals of various densities.

1.2.2 Some definitions

One can define the *size* or *dimension of a particle* by a characteristic length, which best describes its dispersion. For a spherical particle this dimension would be its diameter. For a non-spherical particle, a whole range of different diameters, d_i, can be defined in terms of the lengths of chords joining two points on the surface of particle and going through the centre of gravity. In Table 1.1, only values of the largest diameter, $d_{i\,max}$, of particles are given. Clearly, for a non-spherical particle a quasi-infinite number of such diameters exists and the statistical mean provides the best representation of the dimensions of such a particle. The following mean values are most commonly used:

Arithmetic mean

$$A = \frac{1}{n} \sum_{d_{i\,min}}^{d_{i\,max}} d_i, \tag{1.3}$$

Geometric mean

$$G = \Big(\prod_{d_{i\,min}}^{d_{i\,max}} d_i \Big)^{1/n}, \tag{1.4}$$

Harmonic mean

$$H = \left(\frac{1}{n} \sum_{d_{i\,min}}^{d_{i\,max}} \frac{1}{d_i} \right)^{-1}, \tag{1.5}$$

where n is the number of particles with diameter d_i.

Arithmetic and geometric means are most frequently used and it is easy to see that the values of these means differ and, moreover, that the difference between them decreases as n increases and as the ratio of maximum to minimum diameter decreases. So for $d_{i\,max}/d_{i\,min} = 10$ and $n = 2$, the percentage difference between geometric and arithmetic means, $100\,(1 - G/A)$, is 43% and with the same ratio and $n = 100$ it is 11%. For $d_{i\,max}/d_{i\,min} = 2$ and $n = 2$ this difference is 6%, for the same $d_{i\,max}/d_{i\,min}$ and for $n = 100$, only 2%.

The shape of a particle. Irregular particles of a solid can have different shapes and it becomes necessary to characterize shape in some way. The ratio $d_{i\,max}/d_{i\,min}$ is often used for this purpose. Another frequently used quantity is the "shape factor" — that is, the ratio of the actual volume or external surface area to the volume or surface area resulting from the measured (effective) dimension. Yet another shape factor is the *sphericity* of a particle, that is, the ratio of the surface area of a sphere to that of a particle of a given volume. Clearly, the sphericity of a particle defines its deviation from the spherical shape.

1.2.3 The distribution of particle sizes

Monodisperse or nearly monodisperse systems occur only rarely; most highly dispersed solids are polydisperse systems, containing particles whose dimensions vary over a wide range. To quantitatively describe the size of particles in such systems, *distribution functions* are used which refer to the size, volume or surface area. The mean particle size is not sufficient to describe these systems adequately, and the choice of a suitable mean may be difficult.

The distribution of particle sizes can be given in the form of a table containing the fractions or percentages by number, weight, or volume of particles having a size within a definite interval. Sometimes the cumulative percentage is given, that is, the percentage of particles, with size less than some maximum value. The same data can be also presented in the form of a *histogram*, where the ordinate represents the fraction (or percentage) of particles within size intervals Δx (see Fig. 1.2). Particle size histograms are often used to give a general idea of the distribution of particle size in a given system and to find the continuous distribution functions which provides the best fit to experiment.

The "smoothed" histogram, that is, a histogram with a very large number of finely-spaced intervals ($\Delta x \to 0$) effectively becomes a continuous distribution function, representing the probability density of particles with a given size x_i. Various kinds of distribution functions are described

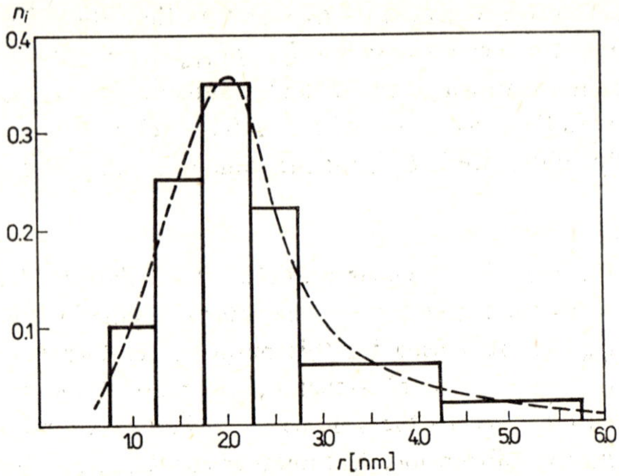

Fig. 1.2 — Histogram of the dimensions of nickel particles in a Ni/Al_2O_3 catalyst (after W. Romanowski [2]).

by mathematical statistics. Assuming that the reader is already familiar with some of the basic notions of statistics, a summary of the properties of distribution functions, which may be useful to describe systems of small particles, is presented below [3].

1.2.3.1 General properties of distribution functions

Since the results of experimental measurements, for instance measurements of particle diameters, are values of a continuous random variable, only the distribution functions of a continuous random variable will be considered. The distribution of a continuous random variable can be expressed in terms of the function describing the probability density, $f(x)$ or the cumulative (integral) distribution function $F(x)$

$$f(x) = \frac{\mathrm{d}}{\mathrm{d}x} F(x), \tag{1.6}$$

where x is the random variable.

Distribution functions contain a number of *distribution parameters*, which characterize the *centre of the distribution*, its *scale* and the *shape* of the distribution curve.

The centre of the distribution is characterized by the following parameters:

1. *Expectation value* (arithmetic mean), $E(x)$

$$E(x) = \int_{-\infty}^{\infty} x f(x) \mathrm{d}x . \tag{1.7}$$

2. The *median* is the value of the random variable z such that

$$\int_{-\infty}^{z} f(x)\,dx = 0.5\,. \tag{1.8}$$

This means that z divides the area under the distribution curve into two halves, or in other words, 50% of the x-values are greater than z.

3. *Mode* — the value of x, for which $f(x)$ is maximum. For the symmetric distributions, e.g. normal (Gaussian) distribution all these three parameters coincide.

The arithmetic mean depends strongly on the minimum and maximum value of the random variable, the median is only weakly dependent, and the mode does not depend on these values at all.

Moments of the distribution. The quantity

$$\mu'_k = E(x^k)$$

is called the k-th moment of the distribution. The *central moment* (k-th moment about the mean) is

$$\mu_k = E[x-\mu'_1]^k = \int_{-\infty}^{\infty} (x-\mu'_1)^k f(x)\,dx\,. \tag{1.9}$$

The first central moment is always zero, $\mu_1 = 0$, since

$$\mu_1 = E(x-\mu'_1) = E(x)-E(\mu_1) = \mu'_1-\mu'_1 = 0\,. \tag{1.10}$$

The second moment about the mean is called *dispersion* *

$$\mu_2 = \sigma^2(x) = E(x-\mu'_1)^2 = \int_{-\infty}^{\infty} (x-\mu')^2 f(x)\,dx\,. \tag{1.11}$$

It can be shown that

$$\sigma^2(x) = \mu'_2-(\mu'_1)^2, \tag{1.12}$$

where σ is the *standard deviation.*

According to the Chebyshev theorem, for an arbitrary distribution having finite mean and dispersion, at least $100\left(1-\frac{1}{k^2}\right)$ per cent of the values of the random variable lie in the interval $\mu \pm k\sigma$.

The third central moment

$$\mu_3 = E(x-\mu'_1)^3$$

* Not to be confused with the physical quantity introduced earlier.

is related to the *skewness* (asymmetry) of the distribution. The measure of the skewness is

$$\alpha_3 = \frac{\mu_3}{\mu_2^{3/2}} . \tag{1.13}$$

The fourth moment

$$\mu_4 = E(x-\mu_1')^4$$

is related to the "steepness" of the distribution peak, called *kurtosis*

$$\alpha_4 = \frac{\mu_4}{\mu_2^2} . \tag{1.14}$$

The parameters denoted by small Greek characters are scale and shape parameters. The complete set of parameters uniquely determines the distribution functions.

1.2.3.2 Some frequently-used distribution functions

The *normal* (*Gaussian*) *distribution* is

$$f(x, \mu, \sigma) = \frac{1}{(2\pi)^{1/2}\sigma} \exp\left[-\frac{(x-\mu)^2}{2\sigma^2}\right], \tag{1.15}$$

where μ is the centre parameter and σ is the scale parameter. This gives the well-known symmetric distribution but it is rarely used in the description of the dimensions of systems of small particles. The cumulative normal distribution function is, according to (1.6),

$$F(x, \mu, \sigma) = \frac{1}{(2\pi)^{1/2}\sigma} \int_{-\infty}^{\infty} \exp\left[-\frac{(x-\mu)^2}{2\sigma^2}\right] dx . \tag{1.16}$$

The *Cauchy distribution function*

$$f(x, \mu, \sigma) = \frac{1}{\sigma\pi}\left[1+\frac{(x-\mu)}{\sigma^2}\right]^{-1} \tag{1.17}$$

is also a symmetric, two-parameter distribution. It is interesting that its moments do not have finite values. It is applied in distributions of random variables with various physical meanings similar to those described for the normal distribution (see also Chapter 3).

The *gamma distribution function* is applied in the description of one-side-bounded random variables. The probability density is

$$f(x,\eta,\lambda) = \frac{\lambda^{\eta}}{\Gamma(\eta)} x^{\eta-1} e^{-\lambda x} \quad \text{with } x \geqslant 0,\ \lambda > 0,\ \eta > 0, \tag{1.18}$$

where η is the shape parameter, λ is a scale parameter and Γ is the well-known gamma function

$$\Gamma(\eta) = \int_0^{\infty} x^{\eta-1} e^{-x} \mathrm{d}x \tag{1.19}$$

(when η is a positive integer, $\Gamma(\eta) = (\eta-1)!$).

The *cumulative gamma distribution function*

$$F(x,\eta,\lambda) = \frac{\lambda^{\eta}}{\Gamma(\eta)} \int_0^{\infty} x^{\eta-1} e^{-x} \mathrm{d}x \tag{1.20}$$

is called an *incomplete gamma function.*

The *beta distribution function* is also a two-parameter, universal distribution function used for the description of random variables bounded over a given range.

The probability density is

$$f(x,\gamma,\eta) = \frac{\Gamma(\gamma+\eta)}{\Gamma(\gamma)\Gamma(\eta)} x^{\gamma-1}(1-x)^{\eta-1} \quad \text{for } 0 \leqslant x \leqslant 1,\ \gamma > 0,\ \eta > 0 \tag{1.21}$$

and zero otherwise.

As seen from Fig. 1.3, the plots of probability density have different shapes, depending on the values of γ and η.

For $\gamma > 1$ and $\eta > 1$, the distribution has one maximum at $x = \gamma - 1/\gamma + \eta - 2$ and is symmetric (Fig. 1.3a) if $\gamma = \eta$ and unsymmetric otherwise (Fig. 1.3b). For cases $\gamma = \eta = 1$ and $\gamma = \eta = 2$, uniform rectangular and parabolic distributions, respectively are obtained.

For $\gamma < 1$ and $\eta < 1$, the distribution is U-shaped and again, is symmetric if $\gamma = \eta$ and unsymmetric if $\gamma \neq \eta$ (Fig. 1.3c).

For $\gamma < 1$ and $\eta > 1$ the distribution is a decreasing function of x (Fig. 1.3d). For the particular case where $\gamma = 1$ and $\eta = 2$, the linear (triangular) distribution is obtained.

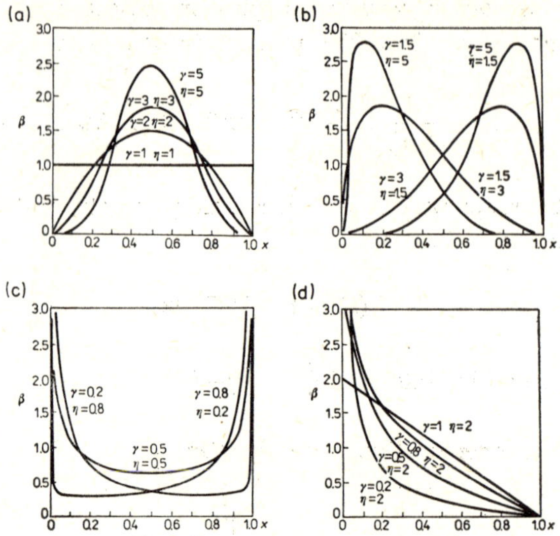

Fig. 1.3 — Beta distribution functions for various values of γ and β.

The cumulative beta distribution function for $0 \leqslant x \leqslant 1$ is

$$F(x, \gamma, \eta) = \frac{\Gamma(\gamma+\eta)}{\Gamma(\gamma)\Gamma(\eta)} \int_0^x x^{\gamma-1}(1-x)^{\eta-1}\,\mathrm{d}x \tag{1.22}$$

and is equal to 1, when $x > 1$. It is the so-called *incomplete beta function.* Values are tabulated for parameter values γ and η in the range 0.5 to 50 in reference [4]. Beta distribution functions have been used to describe the size distribution of particles in suspensions and of glass microspheres [5].

1.2.3.3 Empirical distribution functions

Empirical distribution functions have application in the size analysis of solid particles, formed by various technological processes. For instance, Davies [6] used the function

$$\frac{\mathrm{d}N}{\mathrm{d}D} = KD^{-\gamma}, \tag{1.23}$$

where N is the number of particles with diameter less than D, and K, γ are variable parameters. Denoting N_t as the total number of particles with dimensions smaller than some D_{max} and integrating (1.23) one obtains

$$\frac{N_i}{N_t} = 1 - K\frac{D_i^{1-\gamma} - D_{max}^{1-\gamma}}{(\gamma-1)N_t}. \tag{1.24}$$

The diameter of the smallest particles D_{min} can be obtained by equating the right side of (1.24) to zero

$$D_{min} = [N_t(\gamma-1)/K + D_{max}^{1-\gamma}]^{-1/\gamma-1}. \tag{1.25}$$

The *Weibull distribution function* [7, 8] which is widely used in the analysis of the strength and reliability of materials, is also applicable to small particle dimensions. It is usually applied in the form of a cumulative function F_W, where

$$F_W(x) = 1 - \exp\left[-\frac{(x-\gamma)^{\beta}}{\alpha}\right]. \tag{1.26}$$

Of the three parameters, α and β are scale and shape parameters, respectively and γ is called a *location* or *threshold parameter*.

Differentiation of (1.26) gives the probability density

$$f_W(x) = \frac{\beta}{\alpha}(x-\gamma)^{\beta-1}\exp\left[-\frac{(x-\gamma)^{\beta}}{\alpha}\right]. \tag{1.27}$$

A plot of $f_W(x)$ for various parameter values is given in Fig. 1.4.

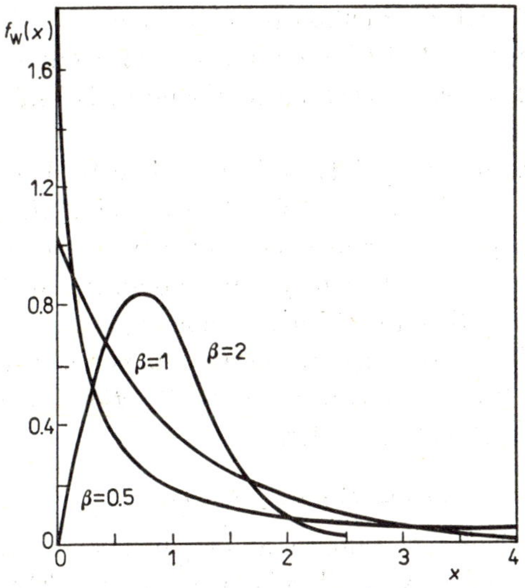

Fig. 1.4 — Weibull distribution functions for various values of β, at $\alpha = 1$ and $\gamma = 0$.

Taking logarithms twice, (1.26) we obtain

$$\ln \ln \frac{1}{1-F_W(x)} = -\ln \alpha + \beta \ln (x-\gamma) , \qquad (1.28)$$

which is a linearized form of the Weibull distribution function. This is plotted in Fig. 1.5 so that the ordinate represents the cumulative percentage of particles with diameters less than x. The quantity γ has a simple

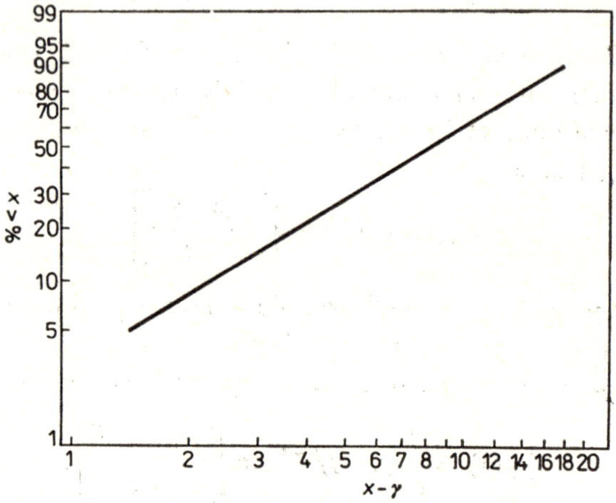

Fig. 1.5 — Representation of the Weibull distribution in linearized form

physical meaning — it represents the smallest particle size present in the system. The Weibull distribution function is quite universal: for $\beta = 1$ the known exponential distribution function is obtained and for $\gamma = 3.25$ the distribution curve coincides with the normal distribution over a wide interval.

Fitting experimental points to the linearized Weibull function requires a choice of γ which is estimated first by successive trials. If the observed values of x are plotted versus cumulative percentage and a straight line is obtained, then $\gamma = 0$. If with $\gamma = 0$ a non-linear plot is obtained, the first trial value is that to which the plot approaches asymptotically. If this value is slightly too large, the curve obtained is concave upward and vice versa. Small corrections in the right direction usually give a correct value of γ and a straight line plot.

1.2.3.4 The log-normal distribution function

The above distribution functions have only limited applications in practice to the statistics of very small solid particles [9]. The familiar normal distribution is in fact, sometimes used to describe the sizes of small particles, despite the fact that a symmetrical distribution of particle sizes occurs only in exceptional cases. On the other hand, distributions with discrete random variables, e.g. the binomial distribution, are cumbersome to apply to sets with a large number of elements. These deficiencies have lead to the exploration of other functions, which more adequately describe the size distribution of small solid particles found in various physical and chemical processes such as fracture, precipitation, reduction, combustion and so on. One of the most important and popular of these is logarithmic-normal (log-normal) distribution function.

Log-normal distribution functions describe particle systems, in which the logarithm of the volume of a particle rather than the volume itself obeys the normal Gaussian distribution. This follows from particle growth theory, assuming growth to be the result of the series of distinct events which enlarge the particle by some fraction of its initial volume [10]. The utility of the log-normal distribution is thus based on a simple law governing the growth of particles. The distribution function is expressed by the formula

$$f_{LN}(x) = \frac{1}{(2\pi)^{1/2}\ln\sigma}\exp\left[-\frac{(\ln x - \ln\bar{x})^2}{2\ln^2\sigma}\right], \tag{1.29}$$

where f_{LN} denotes the normalized distribution function, x is the diameter of a particle and $\bar{x}$ is the median diameter. It is easy to see that this function is analogous to the normal distribution, the quantities σ, x, $\bar{x}$ being converted to their logarithms. From the log-normal distribution functions shown in Fig. 1.6, it can be seen that the distribution is not symmetrical.

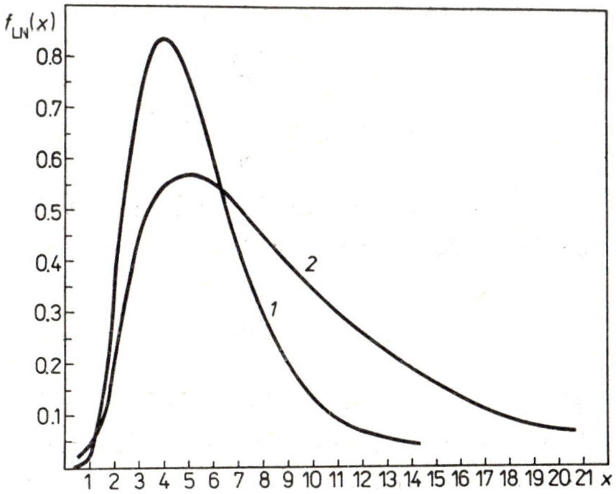

Fig. 1.6 — The plot of the log-normal distribution function $f_{LN}(x)$: *1* — $\bar{x} = 4.1$, $\sigma = 1.6$; *2* — $\bar{x} = 5$, $\sigma = 2$.

Equation (1.29) is valid for particles, whose volumes can be expressed as

$$v = ax^b, \tag{1.30}$$

where a and b are constants, determining the shape of the particle. The parameters $\bar{x}$ (the median) and σ (geometric standard deviation) completely determine the log-normal distribution. These parameters are defined as follows:

$$\ln \bar{x} = \frac{\sum_i \ln x_i}{\sum_i n_i}, \tag{1.31}$$

$$\ln \sigma = \left(\frac{\sum_i n_i (\ln x_i - \ln \bar{x}_i)^2}{\sum_i n_i} \right)^{1/2}, \tag{1.32}$$

where n_i is the fraction of particles in the interval around the central diameter x_i. From equation (1.32) it is clear that the geometrical mean deviation is dimensionless and is always greater than unity. In the limit, when $\ln \sigma \to 0$, the log-normal distribution function approaches the normal distribution.

It has been shown, that $\bar{x}$ is an entirely random parameter, while σ is inversely proportional to the growth (decay) rate of the particles. When the growth process can be taken to be a first order reaction, the parameter σ is a fundamental constant of this process.

The fraction of particles with diameters x in the interval between x_1 and x_2 can be obtained by integration of f_{LN} between corresponding limits

$$F_{x_1 - x_2} = \frac{1}{(2\pi)^{1/2} \ln \sigma} \int_{\ln x_1}^{\ln x_2} f_{LN}(x) \, d(\ln x) . \tag{1.33}$$

It is possible to compare experimental histograms with the corresponding fitted function, using the relation

$$f_{i\,LN} = \frac{\Delta x}{x} f_{LN}(x), \tag{1.34}$$

where $f_{i\,LN}$ is the value of f_{LN} at the point x, Δx is the width of the interval of x in the histogram and $f_{LN}(x)$ is given by (1.29). A fit between the experimental histogram and the theoretical curve may be insufficient to prove that the distribution obtained really is log-normal with parameters $\bar{x}$ and σ. A better procedure is to calculate the fractions, $F_{LN}(x)$, of the total number of particles, having diameters smaller than x (i.e. to obtain cumulative log-normal distribution function $f_{LN}(x)$) and to plot this against $\ln x$. The fraction is given by the formula

$$F_{LN}(x) = \frac{1}{2} + \frac{1}{2}\,\mathrm{erf}\,\frac{\ln(x/\bar{x})}{\sqrt{2}\ln\sigma}, \tag{1.35}$$

where erf is the error function with the argument $\ln(x/\bar{x})/\sqrt{2}\ln\sigma$. The values of this function are tabulated in reference [11]. The values of $F_{LN}(x)$ are marked on the abscissa of Fig. 1.7 as the cumulative percentage of particles with diameter less than x, with $\ln x$ as the ordinate. Divisions on the F axis are symmetrically expanded in both directions from the point $F = 50\%$. In this way, straight line plots of the logarithm of the probability are obtained from which the parameters $\bar{x}$ and σ can be determined. The quantity $\bar{x}$ corresponds by definition to $F = 50\%$; it

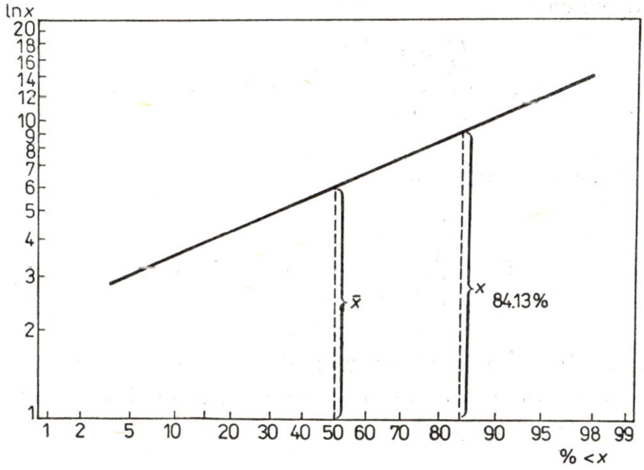

Fig. 1.7 — A logarithmic probability plot.

may be also shown that σ is equal to the ratio $x_{84.13\%}/\bar{x}$ (Fig. 1.7). From the analytical formula (1.29) the moments of the log-normal distribution function [9, 10] can be expressed as follows:

$$\sum_i n_i x^s \sqrt{\sum_i n_i} = \exp(s\ln\bar{x} + \tfrac{1}{2}s^2\ln^2\sigma), \tag{1.36}$$

where s is the order of the moment. These quantities are related to various kinds of mean values, obtained by the measurements of linear dimensions, volumes, surface areas, etc. The corresponding formulae are. the following:

arithmetic mean of the number of particles

$$\bar{x}_a = \frac{\sum_i n_i x_i}{\sum_i n_i} = \exp(\ln \bar{x} + \tfrac{1}{2}\ln^2 \sigma); \tag{1.37}$$

geometric mean of the particle masses

$$\bar{x}_g = \frac{\sum_i n_i x_i^4}{\sum_i n_i x_i^3} = \exp(\ln \bar{x} + 3\ln^2 \sigma); \tag{1.38}$$

mean surface area

$$\bar{x}_{surf} = \left(\frac{\sum_i n_i x_i^2}{\sum_i n_i}\right)^{1/2} = \exp(\ln \bar{x} + \ln^2 \sigma); \tag{1.39}$$

mean volume

$$\bar{x}_{vol} = \left(\frac{\sum_i n_i x_i^3}{\sum_i n_i}\right)^{1/3} = \exp(\ln \bar{x} + 1.5\ln^2 \sigma); \tag{1.40}$$

mean specific surface area

$$\bar{x}_{sp} = \frac{\sum_i n_i x_i^3}{\sum_i n_i x_i^2} = \exp(\ln \bar{x} + 2.5\ln^2 \sigma). \tag{1.41}$$

A useful relationship holds between the modal diameter $\bar{x}_{mod}$, i.e. the diameter corresponding to the maximum value of $f_{LN}(x)$, and the basic parameters of the distribution

$$\bar{x}_{mod} = \exp(\ln \bar{x} - \ln^2 \sigma). \tag{1.42}$$

The various mean values should be carefully distinguished, since they can be significantly different numerically.

In most systems composed of small particles, formed by crystallization or by fracture, the particle dimensions conform to a log-normal distribution. Often, however, some of the particle dimensions present in the sample cannot be described with a single log-normal distribution function. Depending on the particular conditions of particle growth, the distribution may be not uniform, i.e. particles which are present in the sample can belong to different populations. Each population has its own characteristic distribution parameters. The populations are different, when either one or both values of $\bar{x}$ and/or σ differ.

A statistical analysis of non-uniform samples of small particles in photographic emulsions, whose populations conformed to log-normal distribution functions was reported by Kottler [12]. He recommended a numerical method of analysis instead of the simpler, graphical technique. Irani and Callis [9] however worked out a graphical procedure which is equally accurate, but simpler and faster and which lends itself to the analysis of non-uniform systems of particles with log-normal distributions.

To describe the dimensions of particles, other distribution functions are also occasionally applied. We have presented the log-normal distribution in greater detail, since it is most often used in the statistics of small metal particles. This function satisfies the general conditions imposed on distribution functions applied in physical problems [13]; it conforms well to the experimental data, has a comparatively simple mathematical form and is related to the kinetics of formation of particles.

REFERENCES

[1] H. Remy, *Lehrbuch der anorganischen Chemie*, Akademische Verlagsgesselschaft, Leipzig 1955 (8th ed.).

[2] W. Romanowski, *Roczniki Chem.*, **34**, 239 (1960).

[3] G. J. Hahn and S. S. Shapiro, *Statistical Models in Engineering*, John Wiley and Sons Inc., New York 1967.

[4] K. Pearson, *Tables of Incomplete Beta Function*, Cambridge University Press, New York 1934.

[5] G. H. Fricke, D. Rosenthal and G. Walford, *J. Phys. Chem.*, **74**, 1139 (1970); *Anal. Chem.*, **41**, 1866 (1969).

[6] C. N. Davies, *Nature*, **201**, 172 (1964).

[7] W. Weibull, *Nature*, **164**, 1047 (1949); *J. Appl. Mech.*, **18**, 293 (1951).

[8] F. Steiger, *Chem. Technol.*, **1971**, 225.

[9] R. R. Irani and C. F. Callis, *Particle Size: Measurement, Interpretation and Application*, John Wiley and Sons Inc., New Nork 1963.

[10] G. C. Granquist and R. A. Buhrman, *J. Appl. Phys.*, **47**, 2200 (1976).

[11] *Handbook of Chemistry and Physics*, Chemical Rubber Comp., Cleveland 43th and following editions (since 1962), or Hahn and Shapiro *loc. cit.*

[12] F. Kottler, *J. Phys. Chem.*, **56**, 442 (1952).

[13] J. W. J. Fay and J. R. Ashford, *Brit. J. Appl. Phys.*, **11**, 1 (1959).

2

Preparation of Highly Dispersed Metals

2.1 ELEMENTAL METALS

Metal powders, that is, metallic material consisting of separated aggregates of small particles, are usually prepared by chemical processes such as the thermal decomposition of solid or gaseous metallic compounds, or by the reduction of metal ions in solution. Only rarely are highly dispersed metal particles obtained by crystallization from the liquid phase, or by the electrolysis of solutions or molten salts. These methods are normally used to prepare pure elemental metals, but in some cases highly dispersed alloys are also produced for special applications (e.g. as heterogeneous catalysts). Thus methods for the preparation of elemental metals in the form of very fine powders will be described first.

Single grains of metallic powders, however prepared, are usually larger than 1 μm; only rarely do powders have much smaller particle dimensions. Note, however, that the "grains" of a metallic powder are rarely single crystals but are aggregates composed of many contiguous crystallites — that is, particles, which constitute coherently scattering domains when examined by diffraction techniques. Discussion of "particles", or crystallites in metal powders will therefore be in terms of the dimensions of such domains rather than the dimensions of aggregates.

This difference between the dimensions of primary particles and those of their aggregates is especially important when comparing the mean size of particles determined by X-ray methods (Chapter 3) with results obtained by adsorption methods (Chapter 6). The latter techniques measure only the "external" surface accessible for adsorbed molecules and the average size thus obtained may be significantly larger, than the size of the "primary" particles since the surface properties of the aggregate are measured.

2.1.1 Preparation of highly dispersed metals by the decomposition of solids

2.1.1.1 Reduction of metal compounds by gaseous reducing agents

Preparative methods for highly dispersed metals by the thermal decomposition of certain salts of organic acids have been known for some time. Often formates and oxalates are used for this purpose, since they undergo decomposition at comparatively low temperatures, e.g.

$$Ni(HCOO)_2 \nearrow Ni + CO_2 + H_2 \qquad (2.1a)$$

$$Ni(HCOO)_2 \searrow Ni + CO + CO_2 + H_2O \qquad (2.1b)$$

or

$$Ni(COO)_2 \rightarrow Ni + 2CO_2 . \qquad (2.2)$$

Recent investigations of the decomposition of hydrated nickel formate in vacuum [1], have shown that the dehydration of this salt at the temperature of 343–363 K is followed by decomposition at 453–473 K, according to both equations (2.1a) and (2.1b). After reaching 50% decomposition (2.1b) is the predominant reaction.

Very fine powders of iron, cobalt and copper can be obtained in an analogous way. Such powders, with grain diameters below 1 μm are used in the production of special magnetic materials by powder metallurgy techniques [2]. The highly dispersed nickel obtained in this way also finds application as a catalyst in the hydrogenation ("hardening") of vegetable oils to solid fats [3].

Since thermal decomposition reactions may be accompanied by side reactions which lead to the formation of oxides, the decomposition is usually carried out in an atmosphere of hydrogen, thus reducing any oxides to the metal.

Reduction of oxides to metals by gaseous reducing agents such as hydrogen or carbon monoxide, or by solids (carbon, alkali metals) provides a convenient route to the metal. If very highly dispersed metals are required, preference is given to gaseous reduction, since the temperature is then much lower and any subsequent grain growth will be minimized. Alternatively, it may be possible to employ a highly dispersed inactive solid as a diluent, for the same purpose.

Formation of the metal by reduction of the oxide by hydrogen is governed by the equilibrium constant of the reaction, e.g.

$$MeO + H_2 \rightleftharpoons Me + H_2O \quad (2.3)$$

which is expressed by the ratio

$$K_p = \frac{p_{H_2O}}{p_{H_2}}, \quad (2.4)$$

where K_p is the equilibrium constant, and p_{H_2O}, p_{H_2} are partial pressures of hydrogen and water vapour, respectively. In Table 2.1, values of this constant for the reduction of some metal oxides to the metal are given.

Table 2.1 EQUILIBRIUM CONSTANT VALUES $K_p(= p_{H_2O}/p_{H_2})$ FOR THE REDUCTION OF VARIOUS OXIDES BY HYDROGEN AT 673 K

Oxide	K_p	Oxide	K_p	Oxide	K_p
Ag_2O	3×10^{17}	Cu_2O	2×10^{6}	FeO	1×10^{-1}
PdO	1×10^{14}	ReO_3	1×10^{5}	WO_3	1×10^{-1}
IrO_2	1×10^{13}	ReO_2	1×10^{4}	MoO_2	2×10^{-2}
RhO	1×10^{13}	NiO	5×10^{2}	Cr_2O_3	3×10^{-9}
RuO_2	1×10^{12}	CoO	5×10^{1}	MnO	2×10^{-10}
CuO	2×10^{8}	MoO_3	4×10^{1}	TiO	2×10^{-19}
		Fe_2O_3	7×10^{-1}		

During reduction carried out at ambient pressure the partial pressure of hydrogen in the flow apparatus is practically equal to 101 kPa, and the partial pressure of the water vapour formed can be estimated as approximately 0.133 Pa, so that $p_{H_2O}/p_{H_2} \approx 10^{-6}$. A reaction will proceed when this ratio is smaller than the equilibrium constant K_p and as seen from Table 2.1, at 673 K this is the case for all the oxides listed, except the last three.

Preparation of highly dispersed metals is also possible by reduction of their solid chlorides by hydrogen

$$MeCl_2 + H_2 \rightleftharpoons Me + 2HCl\,. \quad (2.5)$$

The extent of this forward reaction is, of course, also given by the value of the equilibrium constant $K_p = p_{HCl}^2/p_{H_2}$ at a given temperature. Values for K_p are generally somewhat higher than those for the corresponding oxides, but this method for obtaining dispersed metals is less often used — evidently because of the higher prices of chlorides, and the corrosive action of hydrogen chloride. The method may however be used to obtain small quantities of the more precious metals, such as those of the platinum group.

Thermodynamic considerations represent necessary but not sufficient conditions for the reaction to occur at a realistic rate. The rate at which thermodynamic equilibrium is attained depends, as with each chemical reaction, on the characteristic value of the activation energy. The principal component of the activation energy is the energy of formation of a nucleus — that is an ensemble of metal atoms, the further growth of which is accompanied by a decrease in free energy. Numerous observations have shown that such nuclei are most often formed at oxide grain boundaries or where the dislocations in the oxides meet the surface or other surface defects. If the activation energy is large, the first stage of the reduction process is very slow. During an "induction period", a sufficiently large population of nuclei is formed and further reduction proceeds by the growth of the nuclei leading to a higher overall rate of reaction. The phenomenon of nucleation at the interface between phases will be considered further in Chapter 5.

The size of crystallites of the metal formed during the reduction process also depends on the particle size of the oxide undergoing reduction and this, in turn, is dependent on the decomposition temperature of the salts or hydroxides from which the oxides were obtained. The lower this latter temperature, the smaller will be the particle size of the oxide and, therefore, of the metal obtained by subsequent reduction. Therefore, it may be preferable to obtain oxides from salts, which decompose at the lowest possible temperatures (in addition to the formates and oxalates already mentioned, nitrates and carbonates can also be used). Another method for obtaining highly dispersed oxides is by the dehydration of hydroxides at relatively low temperatures (below 800 K).

There are however, some processes by which it is possible to obtain a highly dispersed metal by the reduction of a "bulk" (i.e. coarse-grained) oxide. For instance, an inert (irreducible) oxide may be added, which dissolves in the oxide to be reduced and after reduction, the inert oxide accumulates at metal grain boundaries thus preventing further grain growth. An example of this is the preparation of iron catalysts for ammonia and Fischer–Tropsch syntheses by hydrogen reduction of fused magnetite, Fe_3O_4 containing few at % of SiO_2, Al_2O_3 and K_2O. Detailed investigations have shown that after reduction of the magnetite with hydrogen at 653 K, 32–41 nm iron particles are obtained (determined by broadening of X-ray lines, see Chapter 3). After 24 hours at 753 K, the particles have grown to 34–55 nm, clearly demonstrating the effectiveness of the added "structural promoters" in preventing the growth of iron particles [4].

Oxides of platinum metal group (Pt, Rh, Ru, Ir), which can subsequently be reduced by hydrogen to give highly dispersed metals, can be prepared by the method described by Adams [5]. In this process, a chlorine compound of the metal, e.g. chloroplatinic acid H_2PtCl_6 is fused with sodium nitrate at 780 K, giving a highly dispersed oxide

$$\begin{gathered} 6NaNO_3 + H_2PtCl_6 \rightarrow 6NaCl + Pt(NO_3)_4 + 2HNO_3\,, \\ Pt(NO_3)_4 \rightarrow PtO_2 + 4NO_2 + O_2\,. \end{gathered} \tag{2.6}$$

After dissolution of the melt and careful washing of the brown precipitate of PtO_2, the latter is reduced with hydrogen — a reaction which proceeds to completion at room temperature. Highly dispersed metals obtained in this way, have specific surface areas up to 50 $m^2 \cdot g^{-1}$ [6] and are used as catalysts in chemical reactions involving hydrogen (reduction and hydrogenation of organic compounds). Platinum group metals obtained by this method are generally much more highly dispersed than those obtained in other ways, e.g. by hydrogen reduction of chlorides [7].

2.1.1.2 Reduction by solid reagents

Highly dispersed metals can also be prepared by the reduction of their compounds by reducing agents which are solid at room temperature, but which become liquid at the reaction temperature. Historically, the oldest example is the reduction of oxides and chlorides by metals having higher chemical affinity for chlorine or oxygen than the metal to be obtained. For example,

$$TiCl_4 + 4Na \rightarrow Ti + 4NaCl\,, \tag{2.7}$$

$$Cr_2O_3 + 2Al \rightarrow 2Cr + Al_2O_3\,, \tag{2.8}$$

$$ThO_2 + 2Ca \rightarrow Th + 2CaO\,. \tag{2.9}$$

Such reactions are now almost obsolete, both on the technical and the laboratory scale. The processes are uneconomical and technically difficult for large scale production and give metals which are not sufficiently pure for laboratory use.

Some industrial application has been found for a method for preparing iron powder by the reduction of iron oxide by carbon contained in powdered cast iron, by heating appropriate mixtures of partly oxidized, powdered cast iron at a temperature of about 1220 K [2]. Another practical method used in the past for obtaining highly dispersed refractory metals was the reduction of their oxides by solid metal hydrides, for example

$$Cr_2O_3 + 3CaH_2 \rightarrow 2Cr + 3Ca + 3H_2O\,. \tag{2.10}$$

2.1.1.3 Skeletal (Raney) metals

Metals, like nickel, cobalt, iron, copper, silver and rhenium can be obtained in the form of very finely divided powders by leaching their aluminium alloys with concentrated sodium hydroxide solution [8, 9]. The aluminium goes into solution as sodium aluminate, while the other metal forms a suspension of extremely small particles. After separation of the solution, the metals, which are very susceptible to oxidation, are stored under liquid solvents and may be used as catalysts in the reduction of organic compounds. For a number of years, this technique has been mainly used to produce nickel so that our most accurate data relating to the composition, structure and properties refer to Raney nickel.

The chemical composition of nickel–aluminium alloys used for this purpose lies in the range of 40–50 weight % of Ni, corresponding to the intermetallic compounds Ni_2Al_3, and $NiAl_3$, (the compound NiAl does not occur in these alloys [10].) The composition of the leached product depends on the leaching conditions: in addition to nickel, it contains unreacted alloy, a solid solution of aluminium in nickel, hydrated β-alumina and small amounts of hydrogen. As a typical example, the percentage composition of Raney nickel obtained by Fouilloux *et al.* [11] can be represented as 95.4 Ni, 3.5 Al_2O_3, 0.03 Al and 0.5 NiO. Sometimes however, a product with a much higher content of aluminium, both metallic and oxide is obtained.

The specific surface area of skeletal nickel also depends on the leaching conditions and can reach 60–100 $m^2 \cdot g^{-1}$ [12]. Electron-microscopic investigations have shown that nickel crystallites in the Raney-nickel form a sponge-like structure [11, 13] and are themselves most frequently pyramidal [14]. Individual "grains" have dimensions above 100 nm, but single crystallites, of which the former are composed, have dimensions in the broad range of 2.5–15 nm [13]. Skeletal silver, heated *in vacuo* at 440 K has a much larger average particle size (> 600 nm) and correspondingly lower (0.98 $m^2 \cdot g^{-1}$) specific surface area [15]

2.1.2 Reduction of metal ions in solution

Reduction of metal ions in solution with the aid of liquid or gaseous reducing agents proceeds at room temperature (or slightly higher) and produces a metal with extremely small particle in the form of a powder or sponge. Application of this process is possible for metals which do not react with water. The method has a long history and the variety of reducing agents is great, therefore we restrict discussion to the procedures presently used to obtain the most common highly dispersed metals.

2.1.2.1 Gold, silver and the platinum group metals

Highly dispersed gold [16] and silver [15] can be obtained from complex ions in solution by reduction with formaldehyde. To obtain highly dispersed silver, aqueous solutions of glucose and hydrazine hydrate N_2H_5OH may also be used. Precipitation of gold in the form of a very fine powder is possible with the use of hydrogen peroxide and formaldehyde as reducing agents [17]. Similarly, to obtain "platinum black", which consists of an aggregate of very small platinum crystallites, reduction of choloroplatinic acid by formaldehyde in an alkaline medium [18, 19] has been used, at a temperature of about 270 K to prevent a rapid reaction. In the same way, palladium [18] and ruthenium [20] blacks can be obtained. Palladium black, obtained by the reduction of palladium chloride solution by hydrazine or formic acid, and treated with hydrogen at 373 K, has been found to consist of crystallites with an average diameter of 5–45 nm and surface area 5.8–44 $m^2 \cdot g^{-1}$ [21].

Rhenium black can be obtained by the reduction of Re_2O_7, dissolved in dioxane or ethanol with hydrogen under pressure at temperature in the range 390–510 K [22]. The temperature can be lowered to about 330 K, at a hydrogen pressure of 0.4 MPa, when metallic ruthenium, supported on active carbon is used to catalyse the reduction.

An important method for obtaining the platinum group metals, for subsequent application as catalysts, involves reduction in solution by sodium borohydride, $NaBH_4$ [23]. The particles of platinum and ruthenium thus obtained may have particle sizes of 30–50 nm (as measured by X-ray line broadening) and specific surfaces areas in the range 3–12 $m^2 \cdot g^{-1}$. In other investigations [24], the dimensions of platinum black particles were found to be about 10 nm [24].

Colloidal solutions of noble metals. By reduction of noble metal ions in solution it is also possible to obtain colloidal solutions in various solvents.

The interest in such solutions and their properties dates back to the end of the last century, when methods for their preparation were described. Since exact methods for the determination of particle size were not known at that time, characterization of such solutions has only been possible more recently. The development of electron microscopy and other methods has made it possible to select the most efficient recipes for preparing colloidal solutions of noble metals consisting of extremely small particles with a predetermined, uniform size.

R. M. Wilenzick *et al.* [25] obtained such platinum sols by the reduction of H_2PtCl_6 with sodium citrate after dialysing the product to remove ionic reaction products and applying gelatin as a protective colloid to prevent coagulation. The stable solution thus obtained consisted of platinum particles with uniform dimensions (4 nm). If these particles were then used as "seeds" in the same reaction, Wilenzick *et al.* obtained very uniform platinum particles of diameter 10 nm whereas, if tannin was used as a reducing agent, 30 nm particles were obtained. Gold sols with a uniform particle size of 6 nm could be obtained by reduction of chloroauric acid with phosphorus dissolved in ether, while reduction using sodium citrate resulted in particles of 20 nm diameter. Owing to the addition of gelatin as a protective colloid, it was possible to concentrate these solutions and preserve the particles without aggregation. Using this procedure Marzke *et al.* [26] obtained concentrated (29 weight %) solid suspensions of platinum in gelatin with a very narrow particle size distribution of 2.24 ± 0.32 nm (standard deviation). Using sodium citrate as a reducing agent, gold hydrosols of particle size 19.1 ± 2.1 nm were produced and other workers have prepared gold particles of diameter 3.4 nm using solutions of phosphorus in ether [27]. These authors claim that the dialysis and dehydration procedures used to concentrate and purify dilute solutions do not change the particle size distribution.

2.1.2.2 Non-noble metals

Highly dispersed nickel, cobalt and copper are frequently obtained by reduction of their ions in solution. It is possible to use water-soluble reducing agents, such as sodium borohydride, although in this case products containing boron as impurity are obtained.

A more frequently used reducing agent is gaseous hydrogen. At pressures of the order of 5 MPa and temperatures of about 470 K, the aminocomplexes of the above-mentioned elements are reduced to the metallic state. The sizes of the powder grains thus obtained spread over a wide range but a significant fraction lies below 40 μm [2]. Such methods have some significance in hydrometallurgy and powder metallurgy.

Reduction of aqueous ionic solutions has been used for some time in qualitative and quantitative analytical techniques for the detection and determination of transition metals. For example, molecular and nascent hydrogen or stannous chloride may be used as reducing agents. The metallic precipitates thus obtained have a high degree of dispersion — a rather undesirable attribute in this case since it causes difficulties in the subsequent separation by centrifuge or filtration. In order to obtain larger aggregates of metallic particles coagulating agents may be added.

Extraction of metals from ionic solutions by electrolysis is also a reduction process — in this case reduction by electrons. This process usually gives a bulk metal but it is also possible to obtain metals in a highly dispersed state. An interesting application of this method is in the preparation of very fine powders of metallic iron by the electrolysis of ferrous sulphate solution. Iron may be deposited on a mercury cathode, and being insoluble in mercury, a very fine suspension of particles ($\sim$5 nm) is formed. When the electrolytic cell is isolated from vibrations, the iron particles form branched aggregates (dendrites) several tens of nanometres in size. The iron powder collected, after separation of mercury, is a very good material for use in powder metallurgy, especially in the manufacturing of magnetic materials with a large coercive force and *BH* product [28]. Iron-cobalt alloys obtained by this method with particle sizes around 20 nm have been used to fabricate permanent magnets with *BH* products exceeding those of high quality Alnico magnets [29].

2.1.3 Preparation of dispersed metals from their gaseous compounds

The thermal decomposition of volatile metal compounds such as carbonyls, halides and hydrides can also be used as a route to prepare highly dispersed metals. Iron and nickel powders have been obtained from their carbonyls $Fe(CO)_5$ and $Ni(CO)_4$ on a production scale for several decades. These carbonyls are low-boiling liquids whose vapours decompose at temperatures above 400 K to the metals and carbon monoxide,

$$Fe(CO)_5 \rightarrow Fe + 5CO\,. \tag{2.11}$$

The metal powders thus obtained have a high degree of purity and contain only trace quantities of carbon and nitrogen. Iron powders obtained from iron carbonyl have a grain size of 2–15 μm and a spheroidal structure composed of many concentric spherical layers. The external form and dimensions of the grains depend on the decomposition temperature and other factors which influence the concentration of crystalline nuclei. According to some older reports [30] it is possible, by decomposition of carbonyls to obtain metal grains of about 10 nm in diameter. In the most common procedures, the reaction is allowed to proceed until grains of the required size are grown.

Some refractory metals in a highly dispersed form can be obtained by the thermal dissociation of their halides, e.g.

$$ZrI_4 \rightarrow Zr + 4I \qquad (2.12)$$

at a dissociation temperature which is considerably lower than the melting temperature of the metal. This reaction, which forms the basis of the van Arkel–de Boer process, is normally used to obtain bulk metals, by deposition of the metal at high temperature, for example on an incandescent metal wire.

Gaseous hydrides of tin, antimony and lead decompose easily to the metal and hydrogen, a process used to advantage in analytical chemistry. Because of the low melting temperatures of these metals, they are obtained as thin, mirror-like layers rather than powders.

2.1.4 Preparation of highly dispersed alloys

Some of the above methods also lend themselves to the preparation of highly dispersed metal alloys (intermetallic compounds, solid solutions, multiphase alloys). In powder metallurgy, the need to prepare highly dispersed alloys occurs only rarely, since preformed alloy materials can be obtained more conveniently by mechanically mixing powders of the elements followed by a sintering process. In addition to removing porosity and thereby raising the strength, sintering promotes interdiffusion of the elements leading to the formation of the equilibrium phase.

Highly dispersed metallic alloys, prepared at low temperatures are often used in contact catalysis, both in the laboratory and in industry. Recently, therefore the search for new and efficient alloy catalysts has led to the development of methods for preparing such alloys involving techniques which are mostly based on existing preparative methods for highly dispersed elemental metals.

A Cu-Ni alloy catalyst for the hydrogenation of ethylene was prepared by Best and Russel [31] by precipitation of mixed Ni-Cu carbonates from a nitrate solution. The carbonates were then calcined at 670 K followed by reduction in hydrogen at 770 K for 10 hours. Due to the relatively high reduction temperature (needed to facilitate homogenization of the alloy) the resulting powders had a rather low specific surface area of 0.22–0.67 $m^2 \cdot g^{-1}$ and an average particle size (measured by X-ray line broadening) of about 100 nm. Reduction of mixed oxides, obtained in the same way at 520 and 620 K resulted in an alloy powder with a maximum surface area of 1.8 $m^2 \cdot g^{-1}$ at 53.8 weight % Ni [32].

Pt-Pd and Pd-Rh alloys containing 11–94% of palladium and 8–75% Rh respectively, have been obtained by the reduction of mixed chloride solutions by a 5% solution of sodium borohydride, followed by drying at 350 K and hydrogen reduction at 570 K for 3 hours [33]. The Pt-Pd alloys had a specific area of 3.2 to 6.2 $m^2 \cdot g^{-1}$, and 5.9 to 9.6 $m^2 \cdot g^{-1}$ has been obtained for the Pd-Rh alloys.Particle dimensions lay in the range 10–30 nm. Pt-Ru solid solutions with a Ru content up to 60% have been obtained by the reduction of mixed chloride solutions with sodium borohydride or hydrazine hydrate [34]. The specific surface area lay in the range 3–12 $m^2 \cdot g^{-1}$ and the average particle size was 30–50 nm.

A whole series of platinum and palladium alloys: Pt-Rh, Pt-Ru, Pt-Ir, Pt-Pd, Pt-Co, Pt-Fe, Pt-Ni and Pd-Rh, Pd-Ru, Pd-Co, Pd-Ni have been obtained in the form of "blacks" by the method of Adams, described above [35].

Other catalytically active transition metal alloys have been prepared by method similar to those used to obtain Raney metals. Alloys of several elements, in addition to alloys containing aluminium and nickel have been prepared [36–48]. In most cases, there are no accurate measurements of the chemical and phase compositions of these alloys, nor for those obtained by the Adams method. The exceptions are those cases where solid solutions are found (Pt-Ru, and dilute Ni-Re, Ni-Pd, Ni-Pt and Ni-Rh solid solutions).

Metallic alloys dispersed on the surface of non-metallic suports will be considered later in Section 2.3.3 and in Chapter 6.

2.2 PHYSICAL METHODS FOR THE PREPARATION OF HIGHLY DISPERSED METALS

"Physical" methods for the preparation of highly dispersed metals are based, mainly, on evaporation of the heated metal and condensation of its vapour under conditions which limit the growth of crystallites. Frequently used methods are as follows:

(1) evaporation in an electric arc in a liquid or gaseous medium,

(2) evaporation in an inert atmosphere,
(3) evaporation in vacuum,
(4) DC or high frequency sputtering,
(5) "atomization" of liquid metals in a gas stream.

Mechanical crushing (milling, filing), etc. of bulk metals may also be used to achieve a moderately high degree of dispersion.

2.2.1 Evaporation in an electric arc

Evaporation of metals in a direct current electric arc, either in a gas or below the surface of a liquid was used, for the first time by Bredig [49] at the beginning of this century, to produce colloidal solutions of several metals. In order to increase the stability of such fine suspensions (sols), protective colloids were added. This method, further developed and improved by Svedberg [50], is still used, but it has rather limited application, since only small amounts of finely dispersed metals are obtained in the form of fairly dilute solutions. More recently, however, some attention has been paid to metals dispersed in organic solvents, due to the possibility of their application as catalysts, as components of electrically conducting metal–polymer composites and as additives for lubricants. Lunina and Novozhilov [51] have described the preparation of organosols of copper, silver, zinc, cadmium, nickel and of platinum in polar (acetone, methanol, propanol, ethyl acetate, amyl acetate, cyclohexanone) and non-polar (benzene, toluene) organic solvents. The dimensions of the metal particles obtained were mostly in the range 10–40 nm, but much smaller particles (1–1.5 nm) of Pt, Ni and Ag have been prepared using polar solvents. These authors introduced some improvements in the method by applying high frequency alternating current to power the arc, which leads to smaller particles, and to more stable suspensions with higher metal concentrations.

2.2.2 Evaporation of metals in an inert gas atmosphere

In this method, the metal is heated to a temperature (usually above the melting point) at which the vapour pressure is sufficiently high. Particles formed in the gas phase nucleate homogeneously, that is, without any interaction with the solid substrate. Metal atoms in the gas phase collide with atoms of the inert gas (e.g. argon, helium, etc.) losing some of their kinetic energy, thus reducing the effective temperature and leading to supersaturation. Atoms then collide with each other to form nuclei of a liquid phase which unite to form small droplets. These, in turn, are carried by convection and may be deposited on a solid substrate, where they crystallize as very small crystals. Depending on the melting temperature of the metal, crystallization within the gaseous phase is also possible.

This method was explored in the first half of this century [52, 53, 54], and developed further by Japanese research workers [55, 56], so that a relatively large amounts of very small polyhedral metal particles can be prepared with perfectly formed external planes. The apparatus is shown schematically in Fig. 2.1. The properties of the product (particle size and

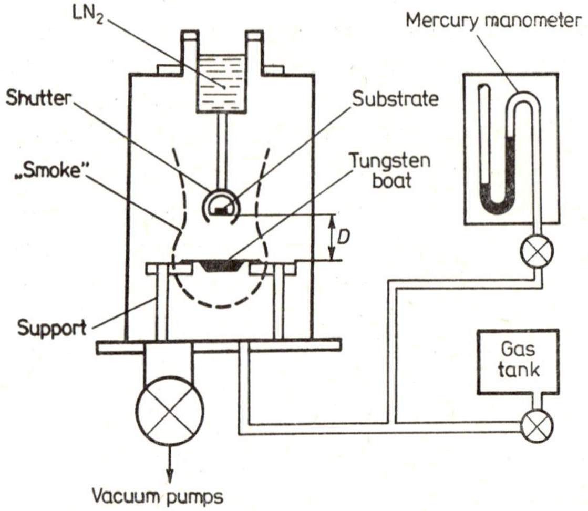

Fig. 2.1 — Schematic diagram of the apparatus for the evaporation of metals in an inert gas atmosphere (after S. Kasukabe *et al.* [66]).

shape) are dependent on the nature of the inert gas and its pressure, on the temperature distribution in the "flame" or "smoke' above the heater, and on the evaporation rate. A typical metal "smoke" plume is bell-shaped at the higher gas pressures, becoming "butterfly"-like at lower pressures (0.266–1.33 kPa) (Fig. 2.2). In the centre of the plume there is a transparent volume corresponding to the uncondensed metal vapour, with larger, polyatomic particles in the outer region. Particles collected at various distances from the heater thus have different sizes and size distributions [57].

The rate of formation of metal particles is largely governed by the vapour pressure of the metal at the temperature of the heater. The vapour pressure of different metals as a function of temperature is given in graphical form in Fig. 2.3.

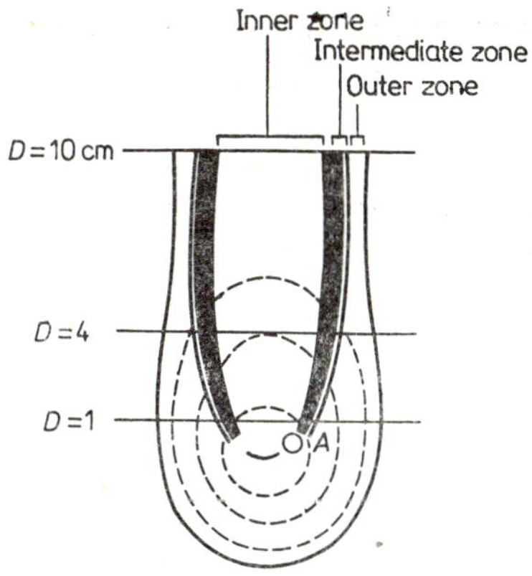

Fig. 2.2 — Three zones of the metal "smoke" in the immediate vicinity of the source (after R. Uyeda [55]).

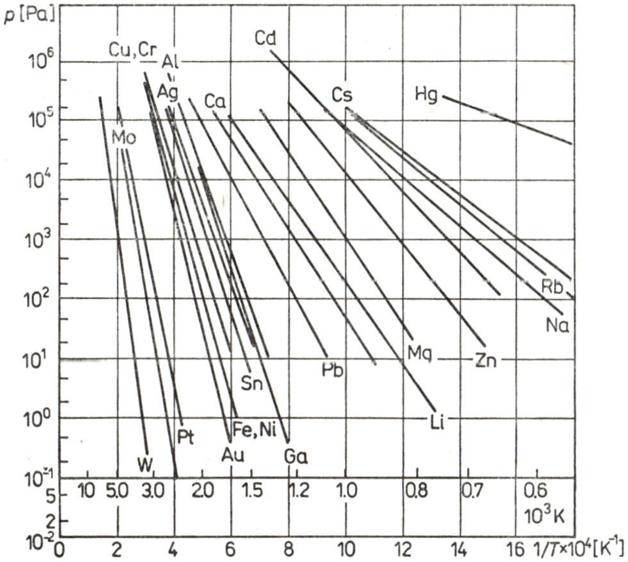

Fig. 2.3 — Vapour pressure — temperature dependence for a number of metals.

Coatings of metal particles, obtained by evaporation and condensation in inert gases are characterized by a loose structure which strongly absorbs visible and infrared radiation — hence "metal blacks" and their original application as radiation absorbers. Owing to the very loose packing of crystallites, the apparent density is very low and the electrical resistivity several orders of magnitude higher than that for a bulk. The conditions

Table 2.2 EVAPORATION OF METALS IN AN INERT GAS ATMOSPHERE

Metal or alloy	Evaporation temperature [K]	Heater	Gas and pressure [kPa]	Average particle diameter [nm]	Ref.
Bi			Air	"Black"	[52]
Bi, Sb			0.03		[53]
Au			N_2, 0.04—0.4	0.04 kPa 1.5—10, 0.4 kPa up to 20	[54]
Au			N_2, 0.4	about 10	[58]
Bi		Tungsten spiral	Air 0.02—0.08	10—40	[59]
Ag-Cu	2153		Ar, He	50—130	[60]
Fe-Co	2373			1.6—8.3	[61]
Mg, Zn, Cd, Al, Fe, Cu, Ag			He	Less than in air	[62]
Be, Mg, Al, Cr, Fe, Ni, Co, Cu, Zn, Ag, Cd			Xe, 0.26—2.13	min. (Ni): 5—60, max. (Mg, Cd, Be, Al): 100—400	[63]
Fe-Co, Fe-Ni Fe-Co-Ni			Ar, 0.4	5—30	[64]
Fe-Co-Ni 63 : 27 : 10		Plasma jet	He, 1.33, 33.25	10—100	[65]
Al	1373, 1573 and 1773	W-boat	He, 1.33, 6.65 and 33.25	10—230	[57]
Mn, Mg Te, Be			Ar, He	Mg > 1 μm Te < 0.5 μm	[66]
Al, Mg, Zn Sn, Cr, Fe Co, Ni, Cu	up to 1873	Graphite crucible W-spiral	Ar, He, Xe up to 19.95	Al, Sn up to 10, Zn up to 1000	[67]

of evaporation and the size of particles obtained are presented for several metals in Table 2.2.

A variant of this method is the condensation of the metal, not in the "static" atmosphere of an inert gas, but in an atomic beam formed by a sonic stream of an inert gas such as argon or helium. A molecular beam, containing metallic atoms and dimers (Fig. 2.4) emerges through

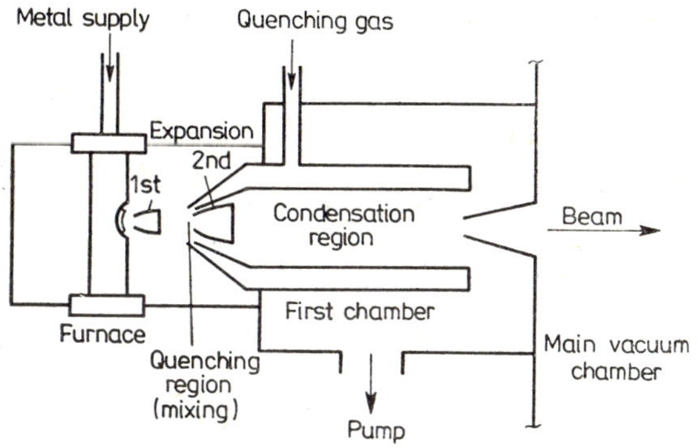

Fig. 2.4 — Apparatus for obtaining metal clusters by a supersonic vapour beam technique (after R.S. Bowles *et al.* [68]).

the orifice into a "quenching zone" where it is mixed with an inert gas and then passes through a second orifice into a condensation zone. By changing the dimensions and the velocity of flow of the molecular beam in the condensation zone (where the pressure is several hundred Pascals) metal clusters of different sizes can be prepared [68]. A second variation of the method consists in replacing the static atmosphere with a flowing inert gas, which carries avay the particles of metal formed and deposits them in a paraffin bath. After solidification of the paraffin well-separated small particles of metal in paraffin matrix can be obtained [68a].

2.2.3 Evaporation of metals in vacuum

Evaporation of metals *in vacuo* and condensation onto a solid substrate has, for many years, been the standard method for the preparation of metals with a high degree of dispersion. Generally, the method is used to obtain continuous thin films of metals; these have wide application and have been the subject of extensive investigation over several decades. Films may be obtained as "island" (discontinuous) structures if evaporation and condensation are terminated when the metal, in the form of isolated particles, covers only a part of substrate. In contrast to evaporation in a gas, particles of the metal are formed by nucleation on the surface of a solid substrate. The properties and the structure of the substrate thus play a substantial role in the mode of formation of metallic nuclei. Growth of the metal particles can also proceed by coalescence of nuclei and small particles migrating on the surface of the support [69, 70]. The mechanism of nucleation and growth of metallic films has been described in many reviews [71–80] and monographs [81–84] and will not be considered here in detail. Although it cannot be assumed that a universal mechanism applies for every metal on any support, the first stage of nucleation always seems to be the adsorption of a metal atom on the surface of the support for a sufficiently long time of adsorption, τ_a, equal to

$$\tau_a = \tau_0 \exp(Q_a/RT)\,. \tag{2.13}$$

Here, Q_a is the heat of adsorption, τ_0 is a constant of the order of 10^{-13} s, R and T have the usual meaning of gas constant and temperature. Once adsorbed on the surface, the atoms change their positions by random displacements (two-dimensional, or surface diffusion) and collide with each other. In this way, nuclei composed of a few atoms are formed, which can grow further by collisions with single migrating atoms and with each other. Another mode of growth is by collision with atoms arriving from the source. When the residence time of atoms on the surface of the support, τ_a, is short, a fraction of the atoms leave the surface by desorption. This model of nucleation and growth of particles on non-metallic substrates holds when the interaction of metal atoms with the substrate, Q_a, is comparatively weak, so that atoms are able to migrate and desorb at the temperature of the substrate. There are, however, cases where this interaction is sufficiently large so that significant migration of atoms and desorption are not possible at moderate temperatures [85]. Nucleation then occurs at specific centres of adsorption on the substrate (immobile adsorption) and by addition of the atoms arriving from the vapour at sites on the substrate at distances smaller than the "capture radius" of the atoms already adsorbed. The capture radius may be a fraction of a nanometre. Atoms of the vapour falling onto the substrate at larger distances from these specific nucleation centres do not desorb, but diffuse into the interior of the substrate. In these cases a low (apparent) value of the sticking coefficient (see Chapter 5) is observed.

It is clear, therefore, that no universal model of the formation and growth of nuclei can be proposed, since the mechanism depends strongly on the nature and magnitude of the specific interactions between the metal atoms and the substrate.

In order to obtain separate small crystallites of the metal on the support, evaporation of the metal must be terminated at a time, when neighbouring particles, while not in contact, nonetheless cover a significant part of the surface area of the support. Depending on the nature of the metal and substrate, the nominal "film thickness" at this stage is 2–20 nm, i.e. it is equivalent to a few tens of atomic layers at most. Such discontinuous metallic films have been obtained and their physical and chemical properties investigated in relation to potential applications as resistors and semitransparent optical filters. They have also been used as laboratory models of metallic catalysts supported on non-metallic substrates.

2.2.4 Cathodic sputtering of metals

Historically, cathodic sputtering preceded vacuum evaporation, and is still used today to obtain thin films of both metallic and non-metallic materials. Sputtering has some practical advantages over evaporation techniques — a greater uniformity of film thickness and composition, the possibility of application to intermetallic and other compounds, but it is generally agreed that contamination is a greater problem and therefore it is less used in fundamental studies.

In this method, the material to be sputtered constitutes the cathode of a glow-discharge diode working at an inert gas pressure of 1.33×10^{-4}–0.133 kPa. The voltage between the cathode (target) and anode is 2–10 keV. Ions of the rare gas (usually argon) produced in the plasma discharge are accelerated by this voltage, hit the target and their impact breaks the bonds between the surface atoms, which evaporate and diffuse through the plasma to the support in the anodic region, where they condense. In addition to single atoms, polyatomic fragments leave the target and condense on the support. The sputtering yield varies considerably, depending on the gas used to produce the plasma and the sputtered metal. This method may be used in several ways — in multielectrode tubes, with a so-called autonomic ion source, at high frequency electric fields — to produce fine-crystalline deposits of platinum metals (which are less prone to contamination by residual gases and impurities dissolved in the metal) and to obtain co-sputtered metal-oxide systems, containing highly dispersed metals in a non-metallic matrix. It also has some importance as a method for fabricating integrated circuits, containing metallic, semiconducting and insulating materials.

2.2.5 Other physical methods

2.2.5.1 Formation of highly dispersed metal in solids by irradiation processes

In many solids containing reducible metal ions; reduction of the latter can be brought about by irradiation and the metal atoms thus formed can act as nuclei for the aggregation of larger particles. A classical example of such phenomena is the photographic process using a solid suspension of silver halides. The so-called photographic latent image probably consists of Ag_n clusters produced by electromagnetic radiation in the visible and ultraviolet or by charged particles, for example electrons. These centres grow to produce silver particles during the development process, which consists in the reduction of Ag^+ ions by chemical reducing agents. Similar systems occur in photosensitive glasses containing, for example, gold ions with cerium and antimony oxides as reducing agents. Irradiation of such glasses with ultraviolet light or other radiation reduces Ce^{4+} to Ce^{3+} which, in turn, can reduce gold ions to neutral atoms at room temperature. These atoms then become centres for the agglomeration of gold atoms obtained by reduction with Sb^{3+} ions on subsequent heating

$$Sb^{3+} + 2Au^+ \rightarrow 2Au + Sb^{5+} . \qquad (2.14)$$

The latter reduction process is comparatively slow and can be interrupted by cooling to room temperature. By controlling the time of reduction, gold particles of various sizes (ranging from 2 to 25 nm in diameter) can be obtained [86].

Silver particles of diameter 1–10 nm can also be obtained in a glassy matrix by dipping an antimony-containing glass into molten AgCl at 870 K [87].

Irradiation of alkali halide crystals by electromagnetic radiation or thermal neutrons produces various lattice defects, for example, the so-called F-centres and interstitial alkali atoms. A neutron dose of over 3×10^{21} m^{-3}, leads to coagulation of Li atoms in LiF at room temperature. Owing to the large concentration of lattice defects, diffusion proceeds without heating the crystal and plate-like lithium particles of maximum diameter about 5 nm can be produced [88, 89]. At neutron doses in excess of 3×10^{22} m^{-3} spherical particles can be formed with diameters in the range 10–100 nm, while the smaller, plate-like particles form even at 77 K.

It has been known for some time that alkali metal particles of "colloidal" dimensions can be produced by a purely chemical process in which alkali metal halides are reduced by metal vapours at elevated temperature.

This leads to the production of F-centres on cooling down to room temperature and coagulation again at higher temperature. Depending on the temperature of coagulation and the initial concentration of F-centres, potassium particles with diameters between 20–100 nm can be obtained by this method [90]. In a similar reaction, silver particles with mean diameters of 33 to 46 nm have been produced in KCl-crystals doped with Ag^+ [91]. During the growth of these particles diffusion of atoms distributed in the alkali halide lattice, represents the rate-limiting step [92, 93].

2.2.5.2 Low-temperature co-condensation of metals with non-metallic matrices (matrix isolation)

This method has been developed chiefly by Klabunde and collaborators to produce microscopic particles of metals to be used as catalysts in reactions involving organic compounds [94–98]. The process consists of

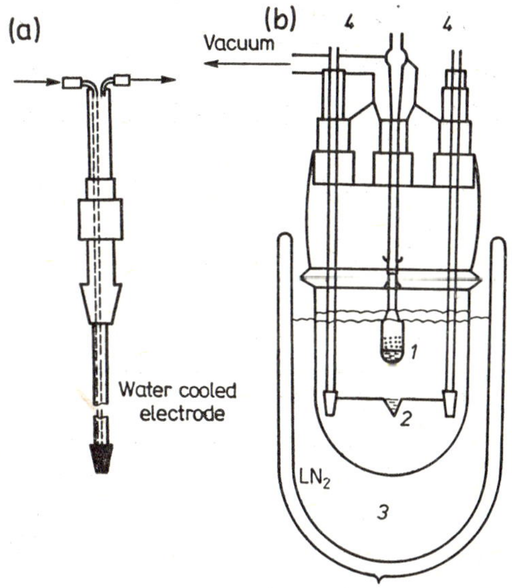

Fig. 2.5 — Apparatus for obtaining co-condensates of metal and an organic solvent (after K. J. Klabunde *et al.* [95]): (a) *1* — solvent evaporator, *2* — metal evaporator, *3* — liquid nitrogen bath, *4* — current electrodes to supply metal evaporator, shown in some detail in (b).

the simultaneous evaporation and condensation of the metal and an organic solvent such as toluene, *n*-pentane or tetrahydrofurane in vacuum at low temperatures. The glass apparatus used for this purpose is shown in Fig. 2.5. After evacuation to about 10^{-2} Pa, a certain amount of the solvent is introduced which then condenses on the walls of the vessel which are cooled to liquid nitrogen temperatures. The tungsten spiral containing the metal is heated and the rate of evaporation of metal and solvent regulated in order to obtain a weight ratio of metal to solvent of the order of a few tenths of one per cent. The solid co-condensate obtained at 77 K contains atomically dispersed metal particles in a solid solvent and may be variously coloured depending on the nature of both species.

After removing the liquid nitrogen bath and allowing the apparatus to warm up to room temperature, the solvent is allowed to melt and can then be removed by pumping. The product is a very fine metal powder, normally in the form of an aggregate with grain size of about 1 μm. The whole process can be presented schematically as

$$\text{Metal atoms (vapour)} + \text{solvent (vapour)} \xrightarrow{77\,\text{K}} \text{Metal atoms in solid solvent} \xrightarrow{\text{melting}} \text{metal} + \text{liquid solvent} \xrightarrow{\text{heating}} \text{suspension of metal in solvent} \xrightarrow{\text{evacuation}} \text{metal powder} + \text{adsorbed solvent} \rightarrow \text{possible pyrolysis of any residual organic material.}$$

The powder obtained after pumping off the majority of the solvent still contains significant quantities of the latter adsorbed on the surface of the metal. In the opinion of the authors [95], even in this state the interaction between the solvent and the metal has the character of a weak chemical bond (complexes or adducts), although the weight ratio of the solvent to a metal such as nickel may be variable: of the order 4% for hexane, 2.5% for toluene and 33% for tetrahydrofurane. Assuming reasonable values for the surface area, covered by one molecule and knowing the overall surface area of the metal, it is possible to estimate the number of monolayers of the solvent on the surface: for nickel this may be one or less to about three. According to X-ray measurements, the primary particles of nickel in an aggregate have dimensions of less than 10 nm [96]. Still smaller nickel crystallites (below 4 nm) have been obtained by introducing a high surface area alumina powder into the condensation vessel before starting to form the toluene matrix [97]. In some nickel/toluene systems, the powders obtained were non-ferromagnetic and even after pyrolysis of the remaining solvent at 773 K, the metal particle size was below 10 nm. In an other publication [98], it is reported that nickel, co-condensed with toluene in the presence of high-area magnesium oxide, had a specific surface area of 19–100 $m^2 \cdot g^{-1}$ and the extent of reduction, that is, the weight ratio of elemental nickel to the total nickel present, was 62–100%.

The organic molecules adsorbed on the surface of metal are believed to inhibit aggregation and recrystallization and also to protect the surface against chemical attack of active gases thus increasing their stability.

Scott and coworkers [99] have obtained small particles of nickel by co-condensation with toluene, sulfur hexafluoride and xenon matrices. The particles of nickel in toluene had the smallest diameters, below 1.6 nm even for large nickel concentrations (up to 13.7 weight %), but organic impurities were always present. Particles within xenon matrices were found to be the most pure. The crystals of nickel were typically 10 nm in diameter and had well-formed external faces.

2.2.5.3 Implantation of metal into oxide substrates

By the bombardment of selected crystallographic planes of α-Al_2O_3 and MgO single crystals with high energy Pt^+ ions at a flux of about $10^{17} cm^{-2}$, metal layers have been formed near the surface of the support [100, 101]. Metal particles with diameters less than 2 nm were found to be buried below the surface of the oxide. Thermal treatment in argon at temperatures up to 1200 K caused the platinum metal to migrate to the surface and the particle size increased to about 20 nm. These materials can be used as catalysts in the hydrogenation of ethylene and the much lower catalytic activity observed in this reaction using a Pt-implanted/MgO catalyst as compared with Pt-implanted/Al_2O_3 suggests a lower surface platinum concentration in the former. The different rates of migration of Pt in these two oxides seems to be connected with the different amount of lattice deformation produced in each oxide by platinum ion implantation. While the surface layer of α-Al_2O_3 became amorphous after bombardment, such damage was not observed in MgO.

2.2.5.4 Mechanical methods

Mechanical methods for the comminution of metals, such as grinding or milling, result in products containing such coarse particles (of the order of microns) that the method is only rarely used, for example in powder metallurgy. The same order of particle size can be obtained by the use of turbulent gas streams as a means of dispersing liquid metals ("atomization"), and this is more often applied in practice in the coating of various materials with metals for anticorrosion or decoration. Both of these methods lead to particle sizes which are too large to produce size-related changes in the physical or chemical properties of metals.

2.3 CHEMICAL METHODS FOR THE PREPARATION OF VERY SMALL METALLIC PARTICLES DEPOSITED WITHIN AND ON THE SURFACE OF SOLIDS

In the preceding paragraphs, methods were presented for the preparation of pure dispersed metals, sometimes containing small additions of a non-metallic phase. Also mentioned were processes for forming metallic particles on non-metallic substrates; for example, discontinuous metal films obtained by evaporation and sputtering. There is however a whole class of materials, containing variable, but usually comparatively small fractions of highly dispersed metals distributed on the surface of non-metallic "supports" or "carriers". These are used as catalysts in various chemical reactions carried out in the laboratory and on an industrial scale. The role of the support, however, is by no means limited to mechanically locating the metal particles, nor even to preventing agglomeration: in many cases the support also participates in chemical and thermal processes involved in the total catalytic reaction. Dilution of the metallic phase by the support may also provide a heat sink so that approximately isothermal reaction conditions are more readily maintained even for strongly exothermic reactions. It is therefore important for the economical application of precious metals catalysts.

Dispersed metals of Groups 1 B and 8 of the periodic table together with their alloys often constitute the metallic phase of typical catalysts. Some typical examples of the preparation of such catalysts and information about the degree of dispersion of the metallic components are given below.

In the preparation of such catalysts it is necessary to ensure a uniform distribution of the precursor compound within the volume of the support, or on its surface. The most frequently used carriers, such as aluminium oxide, silicon dioxide, active carbon, the natural and synthetic alumosilicates, have a large specific surface area (of the order of a few tens to hundreds $m^2 \cdot g^{-1}$) and high porosity, with a part of the pore volume in the form of micropores with a diameter of a few nanometers and less. As a result of capillary forces, uniform *impregnation* of solutions of metal salts and adsorption on the surface becomes easier. Even better distributions of metal ions can sometimes be achieved by *ion exchange*. This is possible for oxides containing a large concentration of surface hydroxyl groups, and particularly, for natural and synthetic zeolites. Zeolites are alumosilicates of the general formula,

$$Me^{n+}_{x/n} \cdot xAlO_2 \cdot ySiO_2 \cdot zH_2O\,,$$

where Me is a metal ion with valency n, or a hydrogen ion. The Me^{n+} ions in zeolites are exchangeable — on contact with a solution containing other metal ions, they go into the solution and their place in the zeolite lattice is taken by ions from the solution. For example

$$\underbrace{Na_{56}(AlO_2)_{56}(SiO_2)_{138} \cdot 264\,H_2O}_{\text{zeolite "Y"}} + 56\,Ag^+ \text{ in solution} \rightarrow$$

$$\rightarrow Ag_{56}(AlO_2)_{56}(SiO_2)_{138} \cdot 264\,H_2O + 56\,Na^+. \qquad (2.15)$$

Similar ion exchange reactions are possible with SiO_2, owing to the presence of surface hydroxyl groups, since hydrogen in OH^- can be exchanged for cations.

Another common method for distributing ions in a carrier is to prepare the carrier and the precursor of the metal simultaneously in one chemical operation, for instance by coprecipitation, e.g

$$Mg^{2+} + Ni^{2+} + (COO)_2^{2-} \rightarrow Mg(Ni)(COO)_2 \text{ (mixed oxalate)} \qquad (2.16)$$

or

$$Al^{3+} + Ni^{2+} + 5OH^- \rightarrow Al(OH)_3 + Ni(OH)_2\,. \qquad (2.17)$$

By heating oxalates (or formates, carbonates or hydroxides) one can obtain mixed oxides, consisting of the supporting oxide together with a more easily reducible oxide which is then converted to a metal by hydrogen reduction.

The preparation of metals supported on SiO_2 and silicates is also possible by direct hydrogen reduction of the corresponding natural or synthetic metal silicates. In the same way metal/zeolite systems can also be obtained by hydrogen reduction of zeolites containing appropriate exchanged ions.

To illustrate these methods, some examples of their use and the characteristics of the products obtained will be given below. The examples constitute, of course, only a small fraction of the number of presently-known supported catalysts.

2.3.1 Iron, cobalt and nickel on carriers

Iron dispersed on a non-metallic support is less often used as a catalyst than iron with a small admixture of stabilizing oxides (as described above). In contrast, cobalt and nickel supported on carriers of diatomaceous earth (SiO_2 of biological origin) — with the addition of thorium oxide and alkali oxides — may be used as a catalyst in the synthesis of organic compounds from the carbon monoxide and hydrogen (Fischer–Tropsch process). Such catalysts are prepared by co-precipitation of nickel (cobalt) and thorium hydroxides in a suspension of diatomaceous earth and subsequent reduction with hydrogen.

Catalysts, containing nickel, supported on aluminium oxide are also widely used in industrial catalysis. They are prepared by co-precipitation of hydroxides or carbonates, followed by heating (calcination) and reduction with hydrogen, or by the impregnation of aluminium oxide with a solution of nickel nitrate again with calcination and reduction.

Catalysts containing 9–27 weight % of nickel, obtained by impregnation of Al_2O_3 with nickel nitrate solution and reduction at 770 K were found to have a metal surface area of 10–19 $m^2 \cdot g^{-1}$ [102]. Calcination at 670 K then leads to a product with a surface area reduced to 4–10 $mg^2 \cdot g^{-1}$. Nickel and cobalt catalysts, supported on magnesium and zinc oxides, obtained by the thermal decomposition of mixed carbonates, oxalates and formates have been described in a series of papers [103–108]. In some cases, [106–108] diameters of the nickel particles (2–20 nm) and their distribution function were determined by magnetic measurements.

Hydrogen reduction of nickel-magnesium chrysotiles (garnierite) $[Ni(Mg)]_3(OH)_4Si_2O_5$ with a "tubular" structure at 520 K gave nickel crystallites about 3 nm in size dispersed in SiO_2 [109]. Increasing the temperature to 770 K, leads to the formation of nickel particles with a larger diameter (about 8 nm). By the reduction of another, "flake-like" form of chrysotile of similar composition, much larger (20 nm) particles of nickel were obtained, which clearly shows the influence of the structure of the carrier on the dispersion of the metal.

Catalysts, consisting of nickel dispersed in synthetic zeolites (A-, X- and Y-) have been obtained by ion exchange of the sodium form of these zeolites with nickel ions, followed by hydrogen reduction [110–113].

Using magnetic methods, it has been established that a significant fraction of the nickel particles possess extremely small (atomic) dimensions. However, during heating *in vacuo* at a temperature of 720 K, these particles agglomerate easily to form much larger ones, having magnetic properties characteristic for bulk nickel.

2.3.2 Supported catalysts containing metals of the platinum group

Catalysts containing platinum metals are used in the chemical industry, mostly in a supported form for economic reasons. Aluminium oxide, alumosilicates, and active carbon are commonly used carriers. Again, deposition of these metals on a carrier is usually by the impregnation of the latter with solutions of salts or chloroacids and, in the case of zeolites, by ion exchange. The particle size distribution of the resulting metal depends on the conditions of thermal treatment before reduction — namely, the temperature of reduction, specific surface area of the support and its porosity.

Determination of the specific surface area of platinum in catalysts containing 0.5 weight % supported on γ-Al_2O_3 and prepared by impregnation with chloroplatinic acid, calcination and hydrogen reduction at 750 K, was carried out by Jaworska and Wrzyszcz [114] by the hydrogen chemisorption method. They established that the specific surface area of platinum depends on the thermal treatment of the catalyst before reduction, and also on the method of preparation of the Al_2O_3. The specific surface area of platinum reached 190 $m^2 \cdot g^{-1}$, corresponding to a mean particle diameter of 1.5 nm.

Catalysts containing 5–10 weight % of ruthenium, rhodium, palladium and iridium have been obtained by impregnation of high area SiO_2 (Cabosil — 300 $m^2 \cdot g^{-1}$), drying and reduction at 720 K [115]. Particle diameters lay in the range 1.4–10.6 nm, the smallest particles being iridium, and the largest palladium. The same authors also investigated the dependence of the particle size of rhodium, obtained as above, on the rhodium content of the catalyst and on the temperature of the thermal treatment after reduction. Changes in the metal content within the range 0.1–1.0 weight % did not influence the particle size (1.1 nm), while heating a silica catalyst containing 5% Rh at 810 K and 1070 K produced an increase in the particle size to 4.1 and 12.7 nm respectively. Unsupported rhodium obtained in the same way, was found to have a particle diameter greater than 250 nm.

Typical Pt/zeolite-Y catalysts, containing 0.25 weight % of Pt, prepared by ion exchange, calcination in air up to 870 K, and reduction with hydrogen at 770 K for 3 hours, were found to have a range of particle diameters of 1.0–7.5 nm [116]. The same catalysts, when only dried at 373 K before reduction consisted mostly of platinum in the form of particles below 1 nm in size.

2.3.3 Metal alloys deposited on carriers

Metal alloys on carriers are prepared by the same chemical methods as single metals (usually impregnation of a carrier with a solution containing ions of the metals in question, followed by drying, calcination and hydrogen reduction). Dispersion of alloy particles prepared in this way generally depends on the factors which determine the properties of single supported metals, in particular the metal content and the surface area of the support. A Pt-Cu alloy on SiO_2 with specific surface area 340 $m^2 \cdot g^{-1}$ and total metal content 5 weight % was found to consist of particles of diameter 5–15 nm [117]. Another supported on SiO_2 with a surface area 700 $m^2 \cdot g^{-1}$ and containing only 0.55 weight % of the alloy, showed a particle size of less than 3 nm.

The degree of dispersion of alloys, both unsupported metals, such as alloy "blacks", and those deposited on carriers, should also depend on the nature of the constituents. The concentration of the component with the higher melting temperature can, for the same preparation conditions, strongly influence the degree of dispersion. This has been shown by Engels *et al.* [118, 119] for the examples of Pt-Ir and Pt-Re alloy powders, as well as Pt/Ir supported on γ-Al_2O_3.

The preparation and properties of Cu-Ni alloy catalysts, supported on SiO_2 [120] and on zeolite [121, 122] have been investigated by magnetic methods. It has been established that the Cu-Ni system reaches thermodynamic equilibrium between the metallic constituents rather slowly, with the formation of a solid solution, at a temperature of 720 K — especially in the case of Cu-Ni on Zeolite-Y. Sinfelt [123] has investigated the catalytic properties of Cu-Ru and Cu-Os alloys supported on silica (with a total metal content of 1 weight %, with atomic ratios of Cu:Ru and Cu:Os in the range 0 to 1). On the basis of their catalytic activities, he concluded that they are miscible (form solid solutions) in the highly dispersed state in contrast to the behaviour of the bulk alloys. The possible changes in miscibility and the accompanying changes of crystal structure for alloys in the highly dispersed state have subsequently been considered by Ruckenstein [124].

Catalytically active Pt-Fe alloys, supported on graphitized carbon black (Graphon) have also been obtained by reduction at 610 and 770 K [125]. The particle diameters of the alloy were 1.5–3.0 nm and no separate phases of Pt and Fe were observed, although platinum and iron constitute a multiphase system in bulk alloys.

REFERENCES

[1] P. G. Fox, J. Ehretsman and C. E. Brown, *J. Catal.*, **20**, 67 (1971).

[2] W. D. Jones, *Fundamental Principles of Powder Metallurgy*, Edward Arnold Publishers, London 1960.

[3] C. L. Thomas, *Catalytic Processes and Proven Catalysts*, Academic Press, London–New York 1970.

[4] F. H. Herbstein and J. Smuts, *J. Catal.*, **2**, 69 (1963).

[5] R. Adams and R. L. Schriner, *J. Am. Chem. Soc.*, **45**, 2171 (1923).

[6] D. W. MacKee, *J. Catal.*, **8**, 240 (1967).

[7] J. H. Sinfelt and D. J. C. Yates, *J. Catal.*, **8**, 348 (1967).

[8] U. S. Patent 1 563 787 (1925), 1 629 191 (1927), 1 915 473 (1933).

[9] H. Adkins and H. R. Billica, *J. Am. Chem. Soc.*, **70**, 695 (1948)

[10] J. Freel, W. J. M. Pieter and R. B. Anderson, *J. Catal.*, **14**, 247 (1970); *ibid.*, **16**, 281 (1970).

[11] P. Fouilloux, G. A. Martin, A. J. Renouprez, B. Moraweck, B. Imelik and M. Prettre, *J. Catal.*, **25**, 212 (1972).

[12] J. Freel, S. D. Robertson and R. B. Anderson, *J. Catal.*, **18**, 243 (1970).

[13] A. Knappvost and K. H. Mader, *Naturwiss.*, **52**, 590 (1965).

[14] J. Yasumura, *Nature*, **173**, 80 (1954).

[15] J. J. F. Scholten, J. A. Konvalinka and F. W. Beekman, *J. Catal.*, **28**, 209 (1973).

[16] R. Zsigmondy, *Z. Physik. Chem.*, **56**, 65 (1906); *ibid.*, **56**, 77 (1906).

[17] W. Trzebiatowski, *Z. Physik. Chem.*, **A 169**, 91 (1934).

[18] R. Wilstätter and E. Waldschmidt, *Ber.*, **54**, 113 (1921).

[19] O. Loew, *Ber.*, **23**, 289 (1890).

[20] M. Kobayashi and T. Shirasaki, *J. Catal.*, **28**, 289 (1973).

[21] H. Kubicka, *J. Catal.*, **5**, 39 (1966).

[22] H. S. Broadbent, G. C. Campbell, W. I. Bartley and J. H. Johnson, *J. Org. Chem.*, **24**, 1847 (1959).

[23] D. W. MacKee and F. J. Norton, *J. Phys. Chem.*, **68**, 481 (1964).

[24] D. W. MacKee *J. Phys. Chem.*, **67**, 841 (1967).

[25] R. M. Wilenzick, D. C. Russel, R. H. Morris and S. W. Marschall, *J. Chem. Phys.*, **47**, 533 (1967).

[26] R. F. Marzke, W. S. Glausinger and M. Bayard, *Solid State Commun.*, **18**, 1025 (1976).

[27] J. Turkevich, P. C. Stevenson and J. Hillier, *Discuss. Faraday Soc.*, **11**, 55 (1951).

[28] F. E. Luborsky, T. O. Paine and L. I. Mendelsohn, *J. Appl. Phys.*, **28**, 344 (1957).

[29] L. I. Mendelsohn, F. E. Luborsky and T. O. Paine, *J. Appl. Phys.*, **26**, 1274 (1955).

[30] C. G. Goetzel, *Treatise on Powder Metallurgy*, Interscience Publishers, London 1949, Vol. 1, p. 51.

[31] R. J. Best and W. W. Russel, *J. Am. Chem. Soc.*, **76**, 838 (1954).
[32] W. K. Hall and P. H. Emmett, *J. Phys. Chem.*, **62**, 816 (1958).
[33] D. W. MacKee and F. J. Norton, *J. Catal.*, **3**, 252 (1964).
[34] D. W. MacKee and F. J. Norton, *J. Phys. Chem.*, **68**, 481 (1964).
[35] G. C. Bond and D. E. Webster, *Ann. New York Acad. Sci.*, **158**, 540 (1969).
[36] T. N. Nalibayev, A. B. Fasman and N. S. Inayatov, *Zh. Fiz. Khim.*, **45**, 383 (1971).
[37] A. B. Fasman, T. Kabiev, D. V. Sokolskii, N. I. Molyukova, A. A. Batkov, I. V. Kirilyus and K. T. Chernousova, *Zh. Fiz. Khim.*, **40**, 114 (1966).
[38] N. I. Molyukova, A. B. Fasman and I. V. Khizhnyak, *Zh. Fiz. Khim.*, **42**, 1673 (1968).
[39] B. F. Petrov, A. B. Fasman and D. V. Sokolskii, *Zh. Fiz. Khim.*, **44**, 3049 (1970).
[40] A. B. Fasman, N. I. Molyukova, T. Kabiev, D. V. Sokolskii and K. T. Chernousova, *Zh. Fiz. Khim.*, **40**, 1758 (1966).
[41] A. B. Fasman, A. Izabekov and B. K. Almashev, *Zh. Fiz. Khim.*, **42**, 903 (1968).
[42] G. A. Pushkareva, A. B. Fasman, Yu. F. Klyuchnikov and I. A. Sapukov, *Zh. Fiz. Khim.*, **46**, 1468 (1972).
[43] A. B. Fasman, A. Izabekov, D. V. Sokolskii, A. A. Presnyakov and K. T. Chernousova, *Zh. Fiz.Khim.*, **40**, 2086 (1966).
[44] A. B. Fasman, D. V. Sokolskii, T. Kabiev, B. K. Almashev and K. T. Chernousova, *Zh. Fiz. Khim.*, **40**, 2212 (1966).
[45] N. I. Molyukova, B. F. Petrov, A. B. Fasman and D. V. Sokolskii, *Zh. Fiz. Khim.*, **41**, 1411 (1967).
[46] A. B. Fasman, T. Kabiev and T. A. Yagudeev, *Zh. Fiz. Khim.*, **41**, 2809 (1967).
[47] A. Izabekov, A. B. Fasman and B. K. Almashev, *Zh. Fiz. Khim.*, **41**, 1890 (1967).
[48] A. B. Fasman, B. K. Almashev, Yu. F. Klyuchnikov and I. A. Sapukov, *Zh. Fiz. Khim.*, **46**, 2559 (1972).
[49] G. Bredig, *Z. Electrochem.*, **4**, 514, 527 (1898); *Z. physik. Chem.*, **32**, 127 (1900).
[50] T. Svedberg, *Kolloid Z.*, **24**, 1 (1919).
[51] M. A. Lunina and Yu. A. Novozhilov, *Koll. Zhurn.*, **31**, 467 (1969).
[52] A. H. Pfund, *Rev. Sci. Instr.*, **1**, 397 (1930).
[53] H. C. Burger and P. H. van Cittert, *Z. Phys.*, **66**, 210 (1930).
[54] L. Harris, D. Jeffries and B. M. Siegel, *J. Appl. Phys.*, **19**, 791 (1948).
[55] R. Uyeda, *J. Cryst. Growth*, **24/25**, 69 (1974); *ibid.*, **45**, 485 (1978).
[56] S. Kashu, M. Nagase, C. Hayashi, R. Uyeda, N. Wada and A. Tasaki, *Jap. Journ. Appl. Phys.*, Suppl. 2, Part 1, 491 (1974).
[57] S. Yatsuya, S. Kasukabe and R. Uyeda, *Jap. Journ. Appl. Phys.*, **12**, 1675 (1973).
[58] L. Harris and J. K. Beasley, *J. Opt. Soc. Am.*, **42**, 134 (1952).
[59] L. Fritzsche, F. Wolf and A. Schaber, *Z. Naturforsch.*, **16a**, 31 (1961).
[60] M. Ya Gen, E. Velichenkova, I. V. Eremina and M. S. Ziskin, *Fiz. Tv. Tela*, **6**, 1622 (1964).
[61] M. Ya Gen, I. V. Eremina and E. A. Fedorova, *Fiz. Met. Metaloved.*, **22**, 721 (1966).
[62] N. Wada, *Jap. Journ. Appl. Phys.*, **6**, 553 (1967).
[63] N. Wada, *Jap. Journ. Phys.*, **7**, 1287 (1968).
[64] K. Kusaka, N. Wada and A. Tasoku, *Jap. Journ. Appl. Phys.*, **8**, 559 (1969).
[65] N. Wada, *Jap. Journ. Appl. Phys.*, **8**, 551 (1969).
[66] S. Kasukabe, S. Yatsuya and R. Uyeda, *Jap. Journ. Appl. Phys.*, **13**, 1714 (1974).
[67] C. G. Granquist and R. A. Buhrman, *J. Appl. Phys.*, **47**, 2200 (1976).

[68] R. S. Bowles, J. J. Kolstad, J. M. Calo and R. P. Andres, *Surface Sci.*, **106**, 117 (1981).
[68a] F. Meier and P. Wyder, *Phys. Rev. Lett.*, **30**, 181 (1973).
[69] W. B. Phillips, E. A. Desloge and J. G. Skofronick, *J. Appl. Phys.*, **39**, 3210 (1967).
[70] R. Kern, A. Masson and J. J. Metois, *Surface Sci.*, **27**, 483 (1971).
[71] J. F. Hamilton, *J. Vac. Sci. Technol.*, **13**, 318 (1976).
[72] J. F. Hamilton and P. C. Logel, *Thin Solid Films*, **16**, 49 (1973); *ibid.*, **23**, 89 (1974).
[73] B. N. Chapman and M. R. Jordan, *J. Phys.*, C, ser. 2, **2**, 1550 (1969).
[74] R. M. Logan, *Thin Solid Films*, **3**, 59 (1969).
[75] I. Markov, *Thin Solid Films*, **6**, 119 (1970).
[76] B. Lewis, *Surf. Sci.*, **21**, 273 (1970).
[77] B. Lewis, *Thin Solid Films*, **7**, 179 (1971).
[78] G. Zinsmeister, *Vakuum Technik*, **22**, 85 (1973).
[79] H. Schmeisser, *Thin Solid Films*, **22**, 83 (1974).
[80] B. Lewis, *Thin Solid Films*, **50**, 233 (1978).
[81] K. L. Chopra, *Thin Film Phenomena*, McGraw-Hill Book Comp., New York 1969.
[82] *Chemisorption and Reactions on Metallic Films*, Ed. J. R. Anderson, Academic Press, London 1971.
[83] W. Romanowski, *Thin Metallic Films*, PWN, Wrocław–Warszawa 1974 (in Polish).
[84] B. Lewis and J. C. Anderson, *Nucleation and Growth of Thin Films*, Academic Press, New York–San Francisco–London 1978.
[85] J. F. Hamilton and P. C. Logel, *Thin Solid Films*, **16**, 49 (1973); *ibid.*, **23**, 89 (1974).
[86] R. D. Maurer, *J. Appl. Phys.*, **29**, 1 (1958).
[87] R. Dupree and M. A. Smithard, *J. Phys.*, **C5**, 408 (1972); *Phys. Status Solidi*, **11a**, 695 (1972).
[88] M. Lambert, Ch. Mazieres and A. Guinier, *J. Phys. Chem. Solids*, **18**, 129 (1961).
[89] M. Lambert and A. Guinier, *J. Phys. Paris*, **24**, 552 (1963).
[90] J. M. Calleja and F. Agullo-Lopez, *Phys. Status Solidi*, **A25**, 473 (1974).
[91] S. C. Jain and N. D. Arora, *J. Phys. Chem. Solids*, **35**, 1231 (1974).
[92] J. M. Calleja and F. Agullo-Lopez, *Phys. Lett.*, **A53**, 317 (1975).
[93] A. E. Hughes and S. C. Jain, *Phys. Lett.*, **A58**, 61 (1976).
[94] K. J. Klabunde, *Acc. Chem. Res.*, **8**, 393 (1975); *Angew. Chem. Int. Engl. Ed.*, **14**, 207 (1975).
[95] K. J. Klabunde, H. F. Efner, T. O. Murdock and R. Ropple, *J. Am. Chem. Soc.*, **98**, 1021 (1976).
[96] K. J. Klabunde, S. C. Davis, H. Hattori and Y. Tanaka, *J. Catal.*, **54**, 254 (1978).
[97] K. J. Klabunde, D. Ralston, R. Zoellner, H. Hattori and Y. Tanaka, *J. Catal.*, **55**, 213 (1978).
[98] K. Matsuo and K. J. Klabunde, *J. Catal.*, **73**, 216 (1982).
[99] B. A. Scott, R. M. Placenik, G. S. Cargill III, T. R. Mc Guire and S. R. Herd, *Inorg. Chem.*, **19**, 1252 (1980).
[100] B. Rabette, M. Che, M. Deane and A. J. Tench, VIe Colloque Franco-Polonaise sur la Catalyse, Compiegne 1977.

[101] P. Rabette, A. R. Gonzalez-Elipe, M. M'Bedi, D. Delafosse, M. Che, A. J. Tench and A. Masson, *Surface Sci.*, **106**, 484 (1981).
[102] C. H. Bartholomew and R. J. Farrauto, *J. Catal.*, **45**, 41 (1976).
[103] H. Dreyer and D. Nehring, *Z. anorg. allg. Chem.*, **300**, 122 (1959).
[104] W. Langenbeck, H. Dreyer, D. Nehring and J. Welker, *Z. anorg. allg. Chem.*, **281**, 90 (1955).
[105] W. Langenbeck and A. Giller, *Z. anorg. allg. Chem.*, **272**, 64 (1953).
[106] W. Romanowski, H. Dreyer and D. Nehring, *Z. anorg. allg. Chem.*, **310**, 286 (1961).
[107] W. Romanowski, *Z. anorg. allg. Chem.*, **351**, 193 (1967).
[108] H. Dreyer, *Z. anorg. allg. Chem.*, **362**, 245 (1968).
[109] C. L. Kibby, F. E. Masoth and H. E. Swift, *J. Catal.*, **42**, 350 (1976).
[110] Pat. U.S.A. 3 013 990 cit. after *Chem. Abstr.*, **56**, 5942 (1962).
[111] W. Romanowski, *Przem. Chem.*, **47**, 741 (1968).
[112] W. Romanowski, *Z. anorg. allg. Chem.*, **351**, 180 (1967).
[113] W. Romanowski, *Roczniki Chem.*, **45**, 427 (1971).
[114] Z. Jaworska-Galas and J. Wrzyszcz, *Chemia Stos.*, **1A**, 105 (1966).
[115] J. H. Sinfelt and D. J. C. Yates, *J. Catal.*, **8**, 82 (1967).
[116] T. Kubo, H. Arrai, H. Tominaga and T. Kunugi, *Bull. Chem. Soc. Japan.*, **45**, 607 (1972).
[117] J. H. Anderson, P. J. Conn and S. G. Brandenberger, *J. Catal.*, **16**, 326 (1970).
[118] S. Engels, M. Wilde and Trân-kim-Thanh, *Z. Chem.*, **17**, 10 (1977).
[119] S. Engels, Trân-kim-Thanh and M. Wilde, *Chem. Tech.*, **27**, 459 (1975).
[120] S. D. Robertson, S. C. Kloet and W. M. H. Sachtler, *J. Catal.*, **39**, 234 (1975).
[121] W. G. Renan, A. H. Ali and G. C. A. Schuit, *J. Catal.*, **20**, 374 (1971).
[122] W. Romanowski, *Reaction Kinetics Catal. Lett.*, **4**, 129 (1976).
[123] J. H. Sinfelt, *J. Catal.*, **29**, 308 (1973).
[124] E. Ruckenstein, *J. Catal.*, **35**, 441 (1974).
[125] C. H. Bartholomew and M. Boudart, *J. Catal.*, **25**, 173 (1972).

3

Physical Properties of Metals which are Sensitive to Dispersion. Application to the Determination of the Degree of Dispersion and the Structure of Small Metallic Particles

3.1 ELECTRICAL PROPERTIES OF DISPERSED METALS

The electrical conductivity of individual grains of metallic powders equals the conductivity of the bulk metal until the particles become so small that the electronic structure is modified. Changes in the electronic structure with particle size will be considered to some extent in Chapter 4; for the moment it suffices to state that such an effect becomes noticeable for particles smaller than 10 nm.

The overall electrical resistance of an ensemble of metal particles larger than 10 nm is mainly determined by the resistance of the interface between them: the lower the intergranular resistance, the smaller is the overall resistance of the sample. The nature of the interface depends mainly on surface area and purity; for loosely packed samples the contact resistance will be larger than for pressed and sintered materials. As mentioned in Chapter 2, the electric resistivity of aggregates of metallic particles obtained by the evaporation and condensation of a metal in a gas atmosphere may be orders of magnitude larger than the resistivity of the bulk metal. Anything that causes a decrease in the porosity of the sample (pressure, annealing), decreases the contact resistance between particles and results in a decrease in the electrical resistivity (increase in conductivity).

3.1.1 Changes in the electrical conductivity during pressing and sintering of metal powders

Changes in the electrical resistance of metallic powders during sintering and pressing have been measured and used as an indicator of the quality of materials obtained by powder metallurgical techniques. This method was first applied by Sauerwald and Kubik [1] in their investigation of the sintering of copper and iron powders, which were pressed and then sintered at temperatures up to 1170 K (Cu) and 1370 K (Fe). As expected, a decrease in the electric resistivity was observed with increasing sintering temperature.

Measurements of the resistivity and of the temperature coefficient of resistance of copper and gold have shown that metal powders with average grain size less than 1 μm, and at pressures of up to 1.5×10^6 kPa at room temperature, have a resistivity which is much higher than that of the bulk metal [2]. Hot pressing at 370–720 K at the same pressure in a hydrogen atmosphere leads to in values of the sample density, resistivity and temperature coefficient which approach those of the bulk metals.

Measurements of the resistivity of nickel powders (obtained by the decomposition of nickel carbonyl) at various pressures and temperatures were carried out by Grube and Schlecht [3]. The resistance of their samples depended strongly on the applied pressure in the lower temperature range (600–1100 K), but approached a value characteristic of bulk nickel at sintering temperatures close to 1600 K, regardless of the pressure applied.

Raub and Platte [4] have prepared a series of alloys of gold with iron, nickel, copper, zinc, cadmium and lead, by sintering appropriate mixtures of powders of these metals at temperatures up to 1270 K. These authors used electrical resistance measurements and other properties, as a criterion of the approach to equilibrium in the alloy phase.

Electrical resistance measurements have also been carried out by Martinet and Tacvorian [5] on powders of copper, iron and silver pressed at 5×10^4 to 5×10^5 kPa. On heating samples previously pressed at room temperature to 470 K, they observed an initial increase of resistance and later a rapid decrease. This has been explained as the reduction by hydrogen of oxide layers present at the surface of powder grains. However, Grube and Schlecht have noticed a similar decrease in the resistance of nickel powders in an argon atmosphere and have interpreted this as a result of the desorption of gases which had previously been adsorbed onto the surface of the metal.

The above examples show that measurements of the electric resistance of metal powders and thus of the intergranular resistance, can be of great importance in the investigation of powder metallurgy processes.

3.1.2 The electric conductivity of metallic island films

Dispersed metals can be prepared in the form of a collection of individual particles which are not in contact with each other. This situation prevails, for instance, in island films consisting of small amounts of metals deposited on both porous non-metallic substrates (active charcoal, silica and other oxides) and on non-porous substrates (glass, mica and the alkali halides). In the case of porous substrates, the metal particles can also be isolated from each other at high metal contents. An example of a thin metallic film with separated islands, deposited on insulating substrate is shown in Fig. 3.1.

The electric properties of ultrathin island metal films have been investigated for many years. Metallic particles in such films are a few tens or hundreds of nanometres in size and the gaps between them are of the

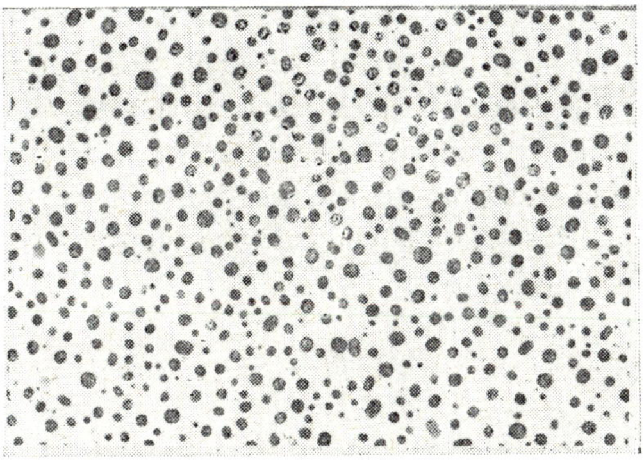

Fig. 3.1 — Electron micrograph of a discontinuous film of platinum supported on silicon dioxide (taken by R. Lamber, Institute of Low Temperature and Structure Research, Polish Academy of Sciences, Wrocław).

same order of magnitude. Transport of electrons from one metal particle to another can take place by following mechanisms:

(i) thermal electron emission,
(ii) field (Schottky) emission,
(iii) quantum mechanical tunnelling.

Thermal emission plays a significant part at the higher temperatures only and in films consisting of islands of the smallest dimensions, separated by comparatively large gaps. The thermal emission current, I, is expressed by the well-known Richardson–Dushman formula

$$I = \frac{4\pi m e^2 k^2 D}{h^3} T^2 \exp\left(-\frac{e\varphi}{kT}\right), \tag{3.1}$$

where k is Boltzmann constant, h is Planck's constant, m and e are the electron mass and charge, respectively, T is the absolute temperature, φ is the barrier height, and D is the "transparence" of the latter. Assuming that for the case of metal islands on a non-metallic substrate, the transfer of electrons from one island to another takes place through the substrate [6], one can equate φ to the difference between the Fermi level of the metal and the bottom of the conduction band of the insulating substrate. When transfer of an electron through the vacuum is assumed, φ is equal to the electron work function.

Purely thermal emission occurs only when the distance between metal particles on the substrate is less than the mean free path of electrons in the insulating substrate; otherwise the electrons emitted into the substrate are captured at local energy levels (traps) in the latter and the conductivity becomes that of the bulk substrate.

Schottky emission can be of importance only in the presence of strong electric fields ($\approx 10^4 \text{V}\cdot\text{cm}^{-1}$) which are capable of lowering the barrier between the metal and substrate. Thus the most probable mechanism responsible for the transport of electrons between metal islands in the absence of high fields and at moderate temperatures can be quantum mechanical tunnelling.

3.1.2.1 Quantum mechanical tunnelling and electronic conduction in thin discontinuous metallic films

The first quantitative theory of electronic conduction in thin discontinuous films was put forward by Neugebauer and Webb [7]. In order to account for the experimentally observed dependence of the resistance with temperature, electric field and particle dimensions, they assumed that the generation of charges in metal particles is thermally activated, and that transfer is realized by tunnelling through the vacuum. According to this theory, the number of charges, n, participating in the conduction process is

$$n = N \exp\left(\frac{\varepsilon_{\text{eff}}}{kT}\right). \tag{3.2}$$

Here, N is the number of metallic particles between the electrodes, ε_{eff} is the effective activation energy-approximately equal to e^2/r (where r is the linear dimension of the particle).

Hence

$$n \sim \frac{1}{r^2} \exp\left(-\frac{e^2/r}{kT}\right). \tag{3.3}$$

When $n \ll N$, i.e. $\varepsilon_{\text{eff}} \gg kT$, the interaction between particles is smallest, and the conductivity, σ, is

$$\sigma \sim P \exp\left(-\frac{e^2/r}{kT}\right). \tag{3.4}$$

P is the electron transfer function, exponentially dependent on $1/T$ over a wide temperature range and the resistance becomes independent of voltage, at least for small applied voltages (ohmic behaviour). The activation energy, however, should depend on particle dimensions. These features of the electrical conductivity have been confirmed experimentally with island films of platinum, gold and nickel. They are also found in the theoretical work of Hartman [8], who considered a model of an island film consisting of a series of equal potential boxes of width a, corresponding to the dimensions of the particles and separated by barriers of width b — equal to the interparticle distances. The activation energy for the generation of conduction electrons was defined as the difference between the energy of the highest occupied electronic state of the particle, E_n, and that of the lowest unoccupied one, E_{n+1}

$$E_a = \Delta E_n = E_{n+1} - E_n = \frac{h^2\pi^2}{2ma^2}(2n+1). \tag{3.5}$$

The conductivity, σ, according to this theory is of the form

$$\sigma = \sigma_0 \exp\left(-\frac{E}{kT}\right), \tag{3.6}$$

where

$$\sigma_0 = \frac{8\pi me^2}{h^3}\varepsilon(a+b)P(E_{n+1}) \tag{3.7}$$

and $P(E_{n+1})$ is the transition coefficient which is constant over the energy interval $E_{n+1} \pm \varepsilon$.

From this theory, the dependence of the activation energy E_n on the particle size was also obtained

$$a = \frac{h^2\pi^2}{4m\Delta E_n d}\left[1+\left(1+\frac{8m\Delta E_n d^2}{h^2\pi^2}\right)^{1/2}\right], \tag{3.8}$$

where d is the distance between neighbouring atoms in the metal.

For sufficiently large particles, $a \gg d$, equation (3.8) becomes

$$a \approx \frac{h^2\pi^2}{2m\Delta E_n d}. \tag{3.9}$$

This simple relation has been confirmed by the experimental data of Neugebauer and Webb [7], where activation energies in the range of 0.04 to 0.64 eV were obtained corresponding to particle sizes between 30 and 2.4 nm, respectively.

It has been suggested by Hill [9, 10] that electron tunnelling can occur through the substrate. Based on this assumption, he developed another theory of conduction in discontinuous metal films [6] and obtained agreement with his own experimental data on films of platinum, gold, silver and chromium, evaporated in ultra-high vacuum.

Four types of island films can be distinguished, for which formulae for the current density have been established within the framework of Hill's model:

type 1: small particles, small interparticle distances,

type 2: small particles, large interparticle distances with thermal emission,

type 3: large particles, small interparticle distances,

type 4: large particles, large interparticle distances *.

The formulae for the current density are too complex to be discussed here, but it should be mentioned that, both in theory and experiment, cases can be distinguished where contributions to the conductivity

* The size of metal particles and of the gaps between them depend on the physical conditions during deposition, for instance, the nature of the substrate and the substrate temperature, the flux density of the metal vapour, the quality of the vacuum, the presence of electrical and magnetic fields, etc.

involving electron transport through the bulk substrate, for by thermal emission are significant. For type 1 films, the activation energy for conduction was found to be

$$E_a = \frac{e^2(d+2s)}{\varepsilon\varepsilon_0 d(d+s)} . \tag{3.10}$$

Here, d and s are the average particle size and distance between particles, respectively and ε is the dielectric constant of the substrate. Fair agreement between observed and calculated values of the conductivity was obtained.

These two early attempts to quantitatively describe the electrical conduction of island films have not, of course, proved to be complete and entirely satisfactory. They have not removed doubts and uncertainties surrounding conduction mechanisms and the generation of charge carriers. Nor have they covered other experimental observations, such as the generally observed non-linear dependence of the resistance on voltage (non-ohmic behaviour), in these systems. There are more recent reviews of the problem (see, for instance [11]) based on more recent experimental and theoretical work, in which the authors conclude that the problem of electrical conduction in island films is still far from a final satisfactory solution.

There have also been many suggested applications for discontinuous thin films in electronics and optics. A major problem, however, is the limited reproducibility of the physical properties of these films caused by their sensitivity to reactive gases, water vapour and temperature. Nevertheless, discontinuous films are currently the subject of basic investigations in their own right and are sometimes used as simplified models of supported metallic heterogeneous catalysts.

3.2 MAGNETIC PROPERTIES OF SMALL METAL PARTICLES

3.2.1 Ferromagnetic metals

3.2.1.1 Saturation magnetization and magnetic moment of small magnetic particles

Frenkel and Dorfman [12] and Néel [13] have shown that small ferromagnetic crystallites (10 nm and smaller) constitute single domains of undirectional magnetization, since further sub-division of such small particles into a larger number of domains would not be compatible with the minimum free energy condition. Another problem to be defined in connection with the magnetic properties of very small magnetic particles is the value of the saturation magnetization of a particle, as compared with that of the bulk metal. The saturation magnetization of ferromagnetic metals (iron, cobalt, nickel and their alloys) is related to their electronic structure in the crystalline state. One can expect therefore that, if the electronic structure of a small particle is different from that of a bulk metal, then the saturation magnetization will also be different. Thus one can suppose that for particles of dimensions below some critical value, the characteristic (parallel) "ferromagnetic" arrangement of atomic magnetic moments will begin to disappear, resulting in a decrease of magnetization as compared with macroscopic crystals.

A theoretical estimate of the lower size limit for particles with ferromagnetic ordering of atomic moments has been given by Wonsowski [14]. In this derivation of the critical linear dimension, δ, Wonsowski applied the uncertainly principle to the electron momentum p, assuming that the electron moves freely inside each particle. The momentum corresponds to the zero point energy which is, in turn, equal to the exchange energy, $\Delta\varepsilon_0 = kT_C$ where T_C is the Curie temperature. Since

$$p = \frac{\hbar^2}{\delta} \tag{3.11}$$

it follows that

$$\Delta\varepsilon_0 = \frac{\hbar^2}{2m\delta^2} \approx kT_C\,. \tag{3.12}$$

Substituting appropriate values for nickel ($T_C = 630$ K), δ is found to be about 0.9 nm; for iron and cobalt still smaller values of δ are found. Particles of dimension less than 1 nm contain only a few atoms. This is a somewhat surprising result, but recently it has been shown experimentally that ferromagnetism in metallic particles persists to such small sizes.

3.2.1.2 Magnetization of systems containing small ferromagnetic particles

Small, single-domain ferromagnetic particles, containing a few tens or a few hundred atoms have a large total magnetic moment, μ, equal to $n\mu_f$. Here, n is the number of atoms in the particle and μ_f the magnetic moment per atom in the ferromagnetic state. The total magnetic moment μ can amount to several hundred Bohr magnetons in such cases. When these ferromagnetic particles are separated by comparatively large distances, their magnetic moments do not interact strongly and they form a system of magnetic dipoles, whose relative orientation can be randomized by thermal motion even at relatively low temperatures. The overall magnetization of the system is therefore similar to that of a paramagnetic material, expressed by the Langevin equation

$$\frac{M}{M_s} = \coth \frac{\mu H}{kT} - \frac{kT}{\mu H} = L\left(\frac{\mu H}{kT}\right), \tag{3.13}$$

where M and M_s are the magnetization in the field H and saturation magnetization, respectively and L denotes the Langevin function. This equation describes the magnetization curves for small ferromagnetic particles dispersed in a non-ferromagnetic matrix, or supported on a non-magnetic substrate, both of which constitute practically important cases. For $\mu H/kT \ll 1$ (if μ is comparatively small and the temperature reasonably high), the Langevin function can be replaced by the approximation

$$L\left(\frac{\mu H}{kT}\right) = \frac{\mu H}{3kT}. \tag{3.14}$$

This leads to an equation which is identical to the Curie law for paramagnetic materials. From Fig. 3.2 it is evident that this is the case for nickel particles containing 100–1300 atoms, at room temperature. For such particles the gram magnetic susceptibility, $\chi_g = \sigma/H$ (where σ is the specific magnetization per unit mass). The quantity χ_g is thus independent of the field strength H, and the reciprocal susceptibility is a linear function of T, as in paramagnetic materials.

For $\mu H/kT \gg 1$ (low temperatures and/or high fields) the Langevin function can be approximated by the formula

$$L\left(\frac{\mu H}{kT}\right) = 1 - \frac{\mu H}{kT}, \tag{3.15}$$

which represents the part of the magnetization curve (M or σ against H) close to saturation.

A system composed of small ferromagnetic particles, whose magnetization conforms to the Langevin equation is called a *superparamagnetic* or a *collective paramagnetic material* [15]. Both of these terms emphasize the fact that (in contrast to the magnetic moments of single atoms, molecules, or ions) we have to do here with the large moments composed of atomic moments.

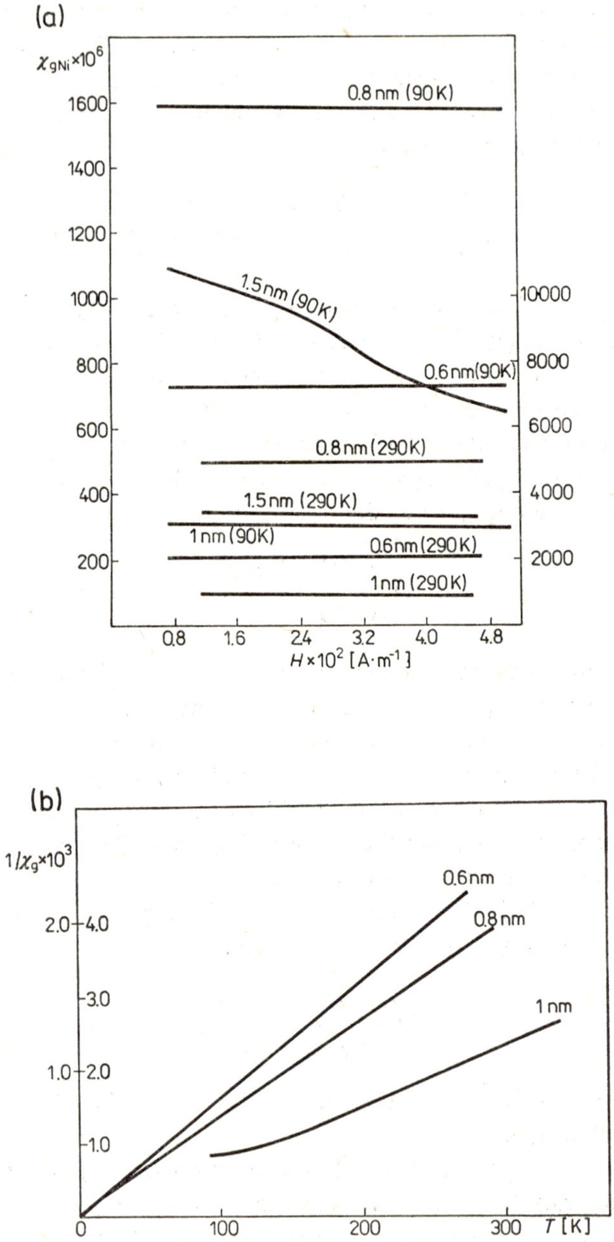

Fig. 3.2 — Calculated magnetic susceptibility of small nickel particles: (a) dependence on the magnetic field strength for 0.6 and 0.8 nm particle diameters (left ordinate), and for 1.0 and 1.5 nm (right ordinate), (b) dependence on temperature.

For polydisperse systems containing ferromagnetic particles of different sizes, the magnetization is expressed by

$$\frac{M}{M_s} = \int_{\mu_1}^{\mu_2} L\left(\frac{\mu H}{kT}\right) F(\mu)\,\mathrm{d}\mu \tag{3.16}$$

or, since $\mu = vI$, where v is the volume of the particle and I is the magnetization per unit volume

$$\frac{M}{M_s} = \int_{v_1}^{v_2} L\left(\frac{vIH}{kT}\right) F(v)\,\mathrm{d}v\,. \tag{3.17}$$

Here, v_1 and v_2 denote the volumes of the smallest and largest particle in the system, respectively, and $F(v)$ is the volume distribution function. The function $F(v)$ can be measured by independent experiments for example, by electron microscopy. In most cases however, $F(v)$ is not known and the magnetic measurements themselves can be used to estimate the distribution in the form of a histogram, as shown by Romanowski [16] and Dreyer [17]. Their method for finding the approximate distribution function is based on the assumption that the integral in equations (3.16) and (3.17) can be represented as a sum

$$\frac{M}{M_s} = \sum_i n_i L\left(\frac{\mu_i H}{kT}\right) \quad \text{with } \sum n_i = 1\,. \tag{3.18}$$

Values of μ_i and n_i are selected so that the experimental and calculated magnetization curves coincide. The fitting can be made in a few trial steps or, alternatively, an exact solution is possible. From the distribution functions determined in this way, the characteristics distribution parameters can be obtained and compared with dispersion characteristics obtained independently.

Most experimental work to date on the magnetic properties of small metallic particles has been carried out with supported nickel catalysts. Investigations have been aimed at the determination of average values of the magnetic moment and volume of metal particles and of the approximate size distribution function in cases where, for various reasons, particle size could not be measured by other methods. The magnetic method spans a wide range of ferromagnetic particle sizes and the technique is comparatively accurate at very small contents of the dispersed ferromagnetic phase in an excess of non-magnetic matrix. In recent work of this kind, magnetic fields of very high intensity (up to 10 Tesla) have been used. This provides the advantage that the saturation magnetization may be approached at ordinary temperatures even with the smallest superparamagnetic nickel particles [18].

3.2.2 Paramagnetic metals

Ferromagnetic metals above their Curie temperatures and certain other transition metals such as palladium, display paramagnetic properties. Since the paramagnetic susceptibility of metals is related to their electronic structure in the solid state, changes in susceptibility should occur as a result of the effect of particle size on electronic structure.

The dependence of the paramagnetic susceptibility of palladium black on the crystallite size (over 5 nm) has been investigated experimentally in some detail by Kubicka [19] who found that particles about 5 nm in size had a susceptibility which was 35% lower than that of the bulk metal.

In contrast to palladium, where the paramagnetic susceptibility originates from unpaired electron spins in each of the atoms constituting the crystal lattice, the majority of solid metals exhibit only "Pauli paramagnetism" due to free (electron gas) conduction electrons. The Pauli paramagnetic susceptibility is about one order of magnitude smaller than the "atomic" susceptibility, and is almost independent of temperature

$$\chi_P = \tfrac{3}{2} N\mu_B^2/\varepsilon_F , \tag{3.19}$$

where N is the number of free electrons per unit mass, μ_B is the Bohr magneton and ε_F is the Fermi energy (of the order of few eV). Since very small metallic particles have only a comparatively small number of electrons occupying states around the Fermi level, and these levels are much more widely spaced than in the bulk metal (see Chapter 4 — electronic structure), there will be unpaired spins in particles containing an odd number of electrons, giving rise to a paramagnetic susceptibility which is different from that of the bulk metal. After Denton *et al.* [20], the expression for the ratio of the susceptibility of small metal particles with an odd number of electrons to that of the bulk metal, is

$$\frac{\chi_{odd}}{\chi_P} = \frac{\delta}{2kT} \quad \text{for } \delta/kT \gg 1 , \tag{3.20}$$

where δ is an average spacing between electron levels near the Fermi level.

The graph of the χ/χ_P against kT/δ (Fig. 3.3) shows that for sufficiently low temperatures, χ/χ_P for the "odd" particles exhibit a susceptibility conforming to the Curie law, while for "even" particles, it increases with temperature. Both susceptibilities converge to a constant value at values of the parameter kT/δ near unity.

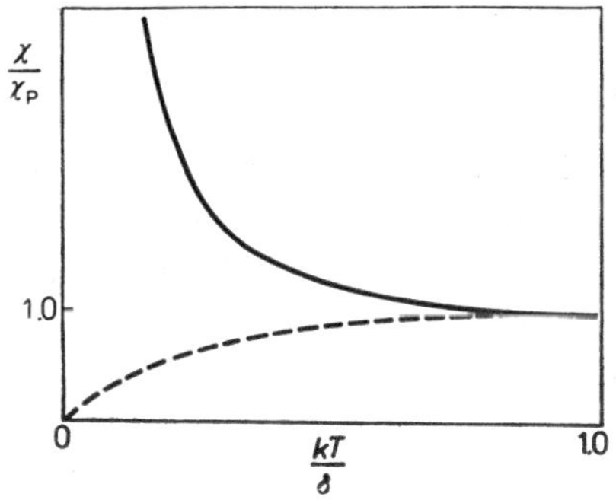

Fig. 3.3 — Magnetic susceptibility of small "odd" (full line) and "even" (dashed line) particles for low values of the parameter kT/δ.

Experimental verification of these theoretically derived relations presents some difficulties, since very small metal particles of uniform size and very accurate susceptibility measurements at low temperatures are needed. Nevertheless, Marzke *et al.* [21] have been able to show that the susceptibility of very small particles of platinum (average diameter of 2.2 nm) changes with temperature below 20 K according to a T^{-1} law and the value of χ corresponds to about one unpaired spin for every two particles.

3.3 INTERACTION OF SMALL PARTICLES WITH ELECTROMAGNETIC RADIATION

3.3.1 Scattering and absorption of light in a medium containing small particles

The interaction of light with an optically inhomogeneous medium, for example one containing small particles with optical properties differing from those of the surrounding matrix, involves two phenomena: absorption and scattering.

Absorption of light is described by the decrease in intensity of the light wave in passing through the substance. Energy is transformed into the internal energy of the absorbing medium (phonons), or into the energy of a secondary radiation such as photoluminescence. Absorption therefore leads to heating of the absorbing substance, or to ionization of atoms and molecules, photochemical reactions, etc.

Absorption of light is described by Lambert's law

$$I = I_0 \exp(-\mu d), \tag{3.21}$$

where I_0 and I are the intensities of the incident and transmitted plane monochromatic waves, respectively, d is the thickness of the absorbing substance, and μ is the absorption coefficient. The value of μ depends on the wavelength and on the nature of absorbing substance. It is related to the imaginary component, $2nk$, of the complex refractive index

$$\mu = 4\pi nk/\lambda, \tag{3.22}$$

where λ is the wavelength in vacuum, n and k are the optical constants of the absorbing substance.

The dependence of μ on the wavelength of the radiation is called the *absorption spectrum*. In contrast to the line spectra of monoatomic gases and vapours and the band spectra of molecules, liquids and solids display quasi-continuous absorption spectra consisting of very broad bands. This is due to the interaction of electron states of atoms at relatively small interatomic distances in the condensed state.

3.3.1.1 Rayleigh scattering

In an inhomogeneous medium, where the refractive index changes from place to place in an irregular fashion, an incident light wave is partially scattered (in all directions). In particular, optical inhomogeneity occurs when small particles are present in a matrix with a different refractive index. When the size of these inhomogeneities (particles) is much smaller than the wavelength of the light (e.g. 0.1λ to 0.2λ), the scattering is called *Rayleigh scattering* or the *Tyndall effect*. In the past, Rayleigh scattering of light was an almost unique method for the detection of small particles and also provided a method for following their motion in transparent liquids (e.g. Brownian movements of colloidal particles). In instruments built for this purpose — called *ultramicroscopes* — the scattered light is observed at an angle of 90° to the incident beam (see Fig. 3.4). In this way it has been possible to establish the presence and follow the trajectories of luminous points, corresponding to single scattering centres (particles) with dimensions below the resolution limit of an optical microscope. Today, this method is used in those cases, for which the application of electron microscopy or X-ray methods (see below) is either impossible or extremely difficult, for example, small particles, dispersed in a glass.

The intensity of the light, I_ϑ, scattered at an angle ϑ to the incident beam by particles of volume v is given by the formula

$$I_\vartheta = I_0 \frac{Nv^2}{r^2\lambda^4} \alpha(1+\cos\vartheta). \tag{3.23}$$

Here, I_0 is the intensity of the incident light, α is a coefficient dependent on the difference between the refraction index of the matrix, n_0 and that

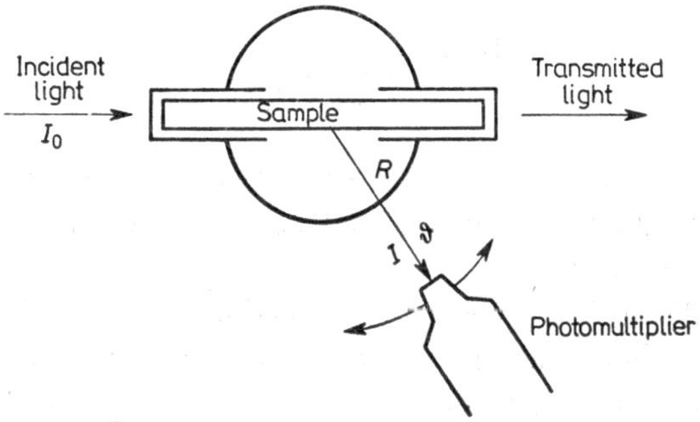

Fig. 3.4 — Measurement of Rayleigh scattering in solids (schematic).

of the particle, n, (when $n-n_0 = 0$, $\alpha = 0$), r is the distance from the scattering volume to the observation point and N is the number of scattering particles per unit volume. The scattered intensity is thus inversely proportional to the fourth power of the wavelength. This is a characteristic feature of Rayleigh scattering, whereas if the size of scattering particles approaches that of the wavelength, the intensity is proportional to λ^{-p}, where $p < 4$.

Experimentally, the so-called Rayleigh coefficient is determined from the relation

$$R = \frac{I_{90}r^2}{I_0 v}, \tag{3.24}$$

where I_{90} is the intensity of light scattered at the angle of 90°. This coefficient is dependent on the optical constants of the system (Mie, 1908)

$$R = \frac{9\pi n_0^4}{2\lambda^4} \frac{(n^2-k^2-n_0^2)^2+4n^2k^2}{(n^2-k^2-2n_0^2)^2+4n^2k^2}, \tag{3.25}$$

where n and k are the optical constants of the particle. Given these constants and experimental measurements of I_0, I and r, v can be determined from equations (3.24) and (3.25), and hence the diameter of the scattering particles.

Equation (3.21) can be re-written to encompass both absorption and scattering, μ being in this case the *extinction coefficient*

$$\mu = NC_{\mathrm{ext}} = N(C_{\mathrm{scatt}}+C_{\mathrm{abs}}), \tag{3.26}$$

where C_{scatt} is the scattering coefficient and C_{abs} is the absorption coefficient.

Mie derived an expression for the extinction coefficient for spherical scattering particles dispersed in a medium of refractive index n_0

$$\mu = \frac{18\pi Nvn_0^3}{\lambda}\cdot\frac{\varepsilon_2}{(\varepsilon_1+2\varepsilon_{\mathrm{m}})^2+\varepsilon_2^2}, \tag{3.27}$$

where $\varepsilon_1+i\varepsilon_2 = n^2$, ε_1 and ε_2 are the real and imaginary parts of the dielectric constant of the particle, and $\varepsilon_{\mathrm{m}} = n_0^2$ is the dielectric constant of the dispersing medium.

From (3.27) it follows that the extinction coefficient is proportional to the particle volume v and to the number of particles per unit volume, N.

Examples of the application of the relations (3.24), (3.25) and (3.27) to the determination of the size of inhomogeneities in glasses and of gold particles in photosensitive glasses are given in papers of Maurer [22, 23] and Smithard and Dupree [24].

The field of the scattered radiation can generally be considered as a superposition of oscillating electric and magnetic multipoles. In the range of visible frequencies and for sufficiently small particles ($d<0.1\lambda$, where d is the particle diameter), contribution from electric dipoles only can be assumed. In this approximation, the results obtained from Mie's theory, contained in equation (3.27) represent plasma oscillations induced in the metallic particles. The induced electric dipole moment, p, is expressed by the formula

$$p = \frac{\pi}{6}d^3\frac{\varepsilon-\varepsilon_{\mathrm{m}}}{\varepsilon+2\varepsilon_{\mathrm{m}}}3\varepsilon_0 E, \tag{3.28}$$

where ε_0 is the dielectric constant of free space, ε is the dielectric constant of the particle and E is the electric field strength.

Plasma oscillations in metallic particles, which give rise to the wonderful colours of metallic sols, are a complex phenomenon and despite almost a century of effort, knowledge is still far from complete. Well-established numerical values for the dielectric and optical constants are not available for all metals, and both classical and quantum approaches to this class of problems (e.g. concerning quantum size effects) are still being published. Concise reviews of the problem can be found in the literature [25].

3.3.2 The broadening of X-ray diffraction lines

The diffraction of X-rays by crystal lattices has been used for many decades to determine the structure, unit cell parameters and the texture, etc. of solids. The Debye–Scherrer powder technique is a standard procedure for the detection and characterization of crystalline phases, including of course, metallic materials.

When the crystallite sizes are below about 100 nm, diffraction lines are broadened in comparison to their normal width, (resulting from various instrumental factors — beam and slit geometry, absorption in the sample, lack of accurate monochromatism, etc.). A quantity which characterizes the broadening of X-ray diffraction line is the half-width, β, of the distribution of scattering angles around some centre value, i.e. the width at half of the maximum intensity of the line, corrected for its instrumental width

$$\beta^2 = B_{obs}^2 - b_{instr}^2, \tag{3.29}$$

where B_{obs} is the total measured half-width of the line in question and b_{instr} is the instrumental half-width. Equation (3.29) is valid for a Gaussian line profile while for a Cauchy profile

$$\beta = B_{obs} - b_{instr}. \tag{3.30}$$

Gaussian and Cauchy profiles of X-ray line intensity are shown in Fig. 3.5.

In practice, the Gaussian profile and relation (3.29) are usually assumed. Instrumental broadening is determined experimentally by the addition to the sample of a standard substance in the form of large crystallites, which gives a sharp intense line in the diffractogram in the proximity of a line of the sample being examined (b_{instr} depends also on the diffraction angle, θ). Having determined the value of β, the average crystallite size, $\bar{d}$, is obtained from the Scherrer formula

$$\bar{d} = \frac{K\lambda}{\beta\cos\theta}, \tag{3.31}$$

where K is the Scherrer constant with the value of about 0.9, λ is the wavelength of the X-radiation and $\bar{d}$ is the mass-average mean crystallite size (see Chapter 1). In Scherrer's original derivation of the formula, the value of K is given as 0.94, whereas in that of Bragg it is 0.89, since it depends slightly on the assumed crystallite shape and symmetry [26].

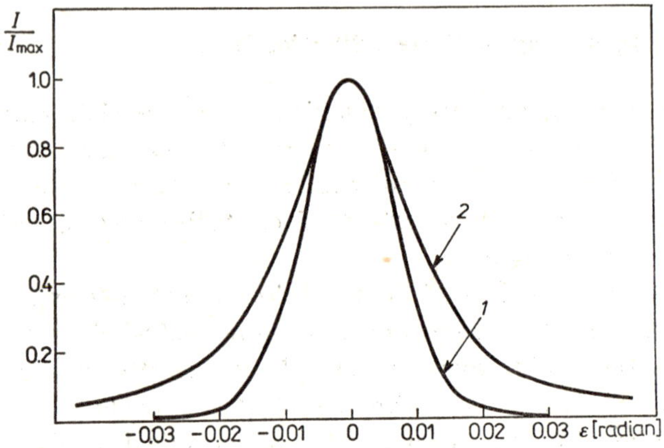

Fig. 3.5 — Gaussian *1* and Cauchy *2* profiles of the intensity of X-ray lines; Gaussian profile: $I = I_{max}\exp(-k^2\varepsilon^2)$; Cauchy profile: $I = I_{max}(1+k^2\varepsilon^2)^{-1}$.

This difference is, however, of little importance, since the method is, at, best, only an approximation. A more general derivation of the equation (3.31) by Stokes and Wilson leads to the effective crystallite size L_{hkl}, which is the volume average of the crystallite size T, normal to the reflecting plane (hkl)

$$L_{hkl} = \frac{1}{v}\int T\,\mathrm{d}v \tag{3.32}$$

with L_{hkl} determined from the relation

$$\beta_i(\theta) = \frac{\lambda}{L_{hkl}\cos\theta}\,. \tag{3.33}$$

This equation is analogous to that of Scherrer with $K = 1$. However, the introduction of L_{hkl} provides the possibility of determining particle dimensions in various directions perpendicular to the diffracting planes (hkl) by an analysis of the broadening of the corresponding lines. This is of considerable importance for particles with large shape anisotropy.

In order to obtain the highest quality line profiles, it is necessary to use diffractometers equipped with counters and a carefully monochromatized X-ray beam and a low angular scan speed. With Cu K_α radiation it is also essential to correct for the doublet $K\alpha_1 - K\alpha_2$ line structure.

It should be remembered when using the Scherrer method that X-ray line broadening can be caused not only by the small sizes of crystallites, but also by the presence of strain and stacking faults in the sample. Thus strong mechanical stresses acting on the specimen during sample preparation (grinding, pressing, filling) should be avoided, or else the stresses removed by appropriate thermal treatment of the sample before X-ray examination. In samples containing strain and stacking faults, an analysis of the line broadening is still possible, although more complicated and less reliable, because the stress and size effects have to be separated. The appropriate methods in this case, proposed by Warren and Averbach [27] and Williamson and Hall [28] are based on a Fourier analysis of broadened line intensities. For less accurate estimates of the contributions of these two effects one can apply a simple method which assumes their additivity and Cauchy line profiles. The stress-induced broadening is then given by the formula

$$\beta_s = K' \tan\theta , \tag{3.34}$$

where K' is a parameter proportional to the stress. The measured line broadening after the correction for instrumental width, is then

$$\beta - \frac{K\lambda}{\bar{d}\cos\theta} + K'\tan\theta . \tag{3.35}$$

Plotting a graph of $\beta \cos\theta$ against $\sin\theta$, $\bar{d}$ is obtained from the intercept with the ordinate.

Methods based on the Scherrer equation, give $\bar{d}$ values to an accuracy of about 30%, the errors being mainly due to the particle size distribution, differences in the shape of the particles. The method can be applied to particles in the size range from about 3 to 50 nm, where determination of the profile of the diffraction line against the background does nct present great difficulties and differences between the instrumental and observed half line widths are large. An additional condition is, of course, that the lines are sufficiently intense and, in the case of the multicomponent samples, that lines of different phases do not overlap.

In the absence of stresses and stacking faults, Fourier analysis of the intensity profiles can be utilized for the determination of the particle size distribution function [29].

Since only standard diffraction equipment is needed, the method of X-ray diffraction line broadening is very often used in the determination of the size of small metallic particles, both in single- and in multiphase systems.

3.3.3 Small angle X-ray scattering (SAXS) method

Determination of the size of small metallic particles, using the information contained in the intensity distribution of X-rays scattered at small angles (up to about 5°) has, to date, been used less often than the Scherrer method. This is probably due to the relative simplicity of the Scherrer technique compared with the theoretically more elaborate small angle scattering method, which in its simpler form can be applied only to "dilute" systems of particles, and whose effectiveness depends, in general, on differences between the scattered intensity from the particles and from the surrounding matrix [30]. The first observations of small angle X-ray scattering date back to the early 1930s and the theory of the phenomenon was worked out later by A. Guinier [31].

The intensity I, of X-rays scattered through a small angle ε by particles of arbitrary shape and orientation, and large compared with the wavelength λ, is given by the relation

$$I = I_0 \exp\left(-\frac{4\pi^2 R^2 \varepsilon^2}{3\lambda^2}\right). \tag{3.36}$$

Here, R is the so-called Guinier radius ("radius of gyration", i.e. the inertia radius with respect to the centre of the electron density), which for a sphere is equal to $(3/5)^{1/2} r$, where r is the geometrical radius of the sphere. For other shapes the Guinier radius can be calculated from geometrical considerations: for an ellipsoid with axes a, a, and na, $R = a((2+n^2)/5)^{1/2}$.

Taking logarithms of (3.36)

$$\ln I = \ln I_0 - \frac{4\pi^2}{3\lambda^2} R^2 \varepsilon^2 \tag{3.37}$$

and by drawing a plot of $\ln I$ against ε^2, R can be obtained from the slope of the straight line (Fig. 3.6).

The theory of small angle X-ray scattering for densely packed particles of arbitrary shape and type of packing was subsequently worked out by G. Porod [32] and O. Kratky [33]. The method of interpretation of the scattering intensity curves (I versus ε^2) given by these authors for polydisperse systems has also proved to be very important. Porod has shown that part of the curve corresponding to high scattering angles, ε, can be expressed by the formula

$$I_{as} = \frac{(\varrho-\varrho_0)^2 S\lambda^4}{8\pi\varepsilon^4}. \tag{3.38}$$

where I_{as} denotes the asymptotic value of the scattered intensity, ϱ and ϱ_0 are the electron densities of the scattering particles and the surrounding medium, respectively, and S is the specific surface area of the particles.

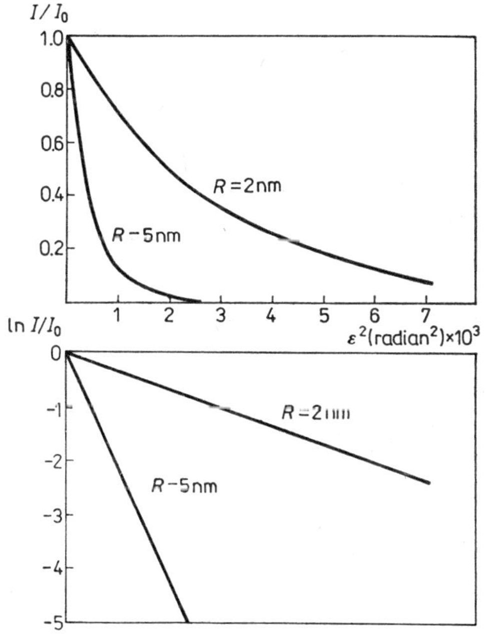

Fig. 3.6 — Graphs of I/I_0 vs. ε^2 and ln I/I_0 vs. ε^2 with assumed values: $\lambda_{CuK_\alpha} = 0.154$ nm, $R = 2$ nm and 5 nm.

This formula is valid for particles of arbitrary shapes and for any packing density. In particular, it can be used to compare the specific surface areas of a metallic phase, dispersed in a medium of small electron density, such as a supported catalyst, by measuring the intensity of scattered radiation for a single value of the angle ε, lying in the range of validity of equation (3.38).

3.3.3.1 Determination of the particle size distribution from small angle scattering

For particles of various sizes present in a polydisperse system, the relation between $\ln I$ and ε^2 is not a straight line. Jellinek *et al.* [34] have shown that by drawing tangents to the curve of $\ln I$ against ε^2 at various values of ε, it is possible to determine the approximate size distribution function by very simple calculations. Bauer [35] has suggested that this function can be obtained by a Fourier transform method from the relation between the scattered intensity and the particle size distribution function, $W(R)$

$$I = c\int W(R)\exp\left(-\frac{4\pi^2R^2\varepsilon^2}{3\lambda^2}\right)\mathrm{d}R\,, \tag{3.39}$$

where $W(R)$ is a distribution function of Guinier radii and c is a constant. Harkness *et al.* [36] have demonstrated the correlation between the distribution parameters $\bar{x}$ and σ (see Chapter 1) and the Guinier and Porod radii (the Porod radius, R_P, being found from the asymptotic part of the $I-\varepsilon^2$ curve). These relations are

$$\ln\bar{x} = \ln R - 1.714\ln R/R_P\,, \tag{3.40}$$

$$(\ln\sigma)^2 = 0.286\ln R/R_P\,. \tag{3.41}$$

Two important examples of the application of small angle X-ray scattering techniques are the determination of the size of Guinier–Preston zones (local inhomogeneities in solid solutions) [37] and the determination of the size of metal particles in supported metal catalysts [38]. Since the particle diameters are generally distributed over a relatively wide range, many researchers have improved the calculation procedure to obtain reasonably accurate values of distribution parameters. In the case of supported catalysts, the presence of pores in the carriers may also be included. Such pores, filled by a gas, constitute a third type of scattering region (apart from the metal and the non-metallic matrix). A procedure for investigating such systems has been proposed by Renouprez and Imelik [39], who have also applied the technique to the determination of the size distribution of platinum particles supported on alumina [40] and on zeolite-Y [41]. In the latter case, the presence of platinum particles as small as 1 nm in diameter was established. Determination of the size of such small particles is beyond the capacity of Scherrer method based on high angle scattering and, is also at the limit of practical electron microscopy.

The instrumental requirements of the technique initially presented some difficulties, since very good collimation of the X-ray beam is required, and diffraction from the edges of the slits must also be carefully eliminated. However, a simple and ingenious "block camera" was constructed by Kratky [42] which removed these technical difficulties so that today, very good apparatus is easily available

3.3.4 Extended X-ray absorption fine structure (EXAFS)

Over the last few years a new and powerful method giving insight into the structure and dimensions of the smallest metallic particles (clusters) with dispersion close to unity, has been widely used. This method is based on absorption of X-rays resulting from the excitation of an electron from core states and the subsequent scattering of the outgoing photo-electron waves by neighbours of the exited atom. Experimentally, the method requires intense X-ray sources and thus became practicable only recently due to the availability of intense synchrotron X-ray radiation.

Atomic absorption spectra of X-rays are known to show sharp discontinuities or "edges" corresponding to treshold energies for excitation of inner core (e.g. K-shell) electrons. If the atoms are bound together in a solid, oscillatory structure is seen in the spectrum: specifically, the absorption coefficient μ against energy E shows fine structure on the high energy side of the absorption edge, extending as far as 1 to 2 keV above

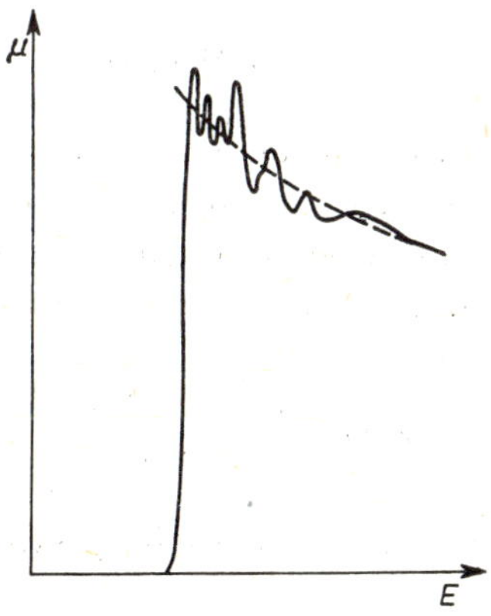

Fig. 3.7 — EXAFS total spectrum-continuous line; the dotted line represents the smooth atomic absorption, characteristic of an isolated atom.

this edge. This fine structure (Fig. 3.7) was first studied by Kronig [43] and a successful theory of the phenomenon was developed by Sayers, Lyttle and Stern in the early 1970s [44, 45].

The phenomenon can be described in terms of two basic processes: first a photoelectron is emitted from the X-ray absorbing atom, giving rise to the major part of the absorption coefficient, and then the photoelectrons are backscattered from atoms surrounding the excited atom. Interference occurs between emitted and backscattered electron waves giving rise to modulations of the absorption coefficient as a function of energy. The final state, ψ_f, close to the nucleus of the excited atom can be described as the superposition of the outgoing "atomic" electron wave ψ_a with waves backscattered from the neighbouring atoms, ψ_{sc}. Thus

$$\psi_f = \psi_a + \psi_{sc} \tag{3.42}$$

and the absorption cross section σ, due to transitions from the initial state, ψ_i is given by

$$\sigma \approx |\langle\psi_i|r|\psi_a\rangle + \langle\psi_i|r|\psi_{sc}\rangle|^2. \tag{3.43}$$

The square of the first term of (3.43) is the atomic cross section which decreases smoothly with energy above threshold. The cross terms yield a contribution which oscillates as a function of final state energy. This oscillation is due to the constructive and destructive interference between the outgoing and backscattered waves at the excited atom. The total cross section, σ_t, can thus be written as

$$\sigma_t(k) = \sigma_a(k)[1+\chi(k)], \tag{3.44}$$

where σ_a is the atomic cross section, k is the wave vector of the outgoing electron, and $\chi(k)$ is the EXAFS modulation due to the presence of the surrounding atoms which can be expressed in terms of atomic $\mu_a(k)$, and total $\mu_t(k)$, absorption coefficients

$$\chi(k) = \frac{\mu_t(k) - \mu_a(k)}{\mu_a(k)}. \tag{3.45}$$

The theoretical expression for $\chi(k)$ derived by Sayers, Stern and Lyttle, is

$$\chi(k) = \frac{1}{k}\sum_j \frac{N_j}{R_j}|f_j(\pi, k)|\sin[2kR_j + \varphi_j(k)]\exp(-2a_j^2k^2 - \gamma R_j), \tag{3.46}$$

where $|f_j(\pi, k)|$ is the backscattering amplitude from each of the N_j neighbouring atoms in the j-th shell a distance R_j from the central (excited) atom. The factor $\exp(-2a_j^2k^2)$ is a Debye–Waller factor which takes account of thermal vibrations, in which a_j^2 is the mean square displacement of atoms from their average positions. The term $-\gamma R_j$ is a photoelectron mean free path, describing the decay of the photoelectron wave. The phase shift, $\varphi_j(k)$ represents the effect of the potentials of absorbing and scattering atoms on the electron wave.

By a Fourier transformation of $\chi(k)$, it is possible to obtain a radial structure function $|F(R)|$ with peaks corresponding to atomic shell radii at distances R_j from a given central atom.

As a structural technique, the EXAFS method has many advantages. Since the X-ray absorption is specific for a given element, it is possible to analyse separately the environment of each kind of atom in a multi-component system. Due to its sensitivity to the immediate neighbourhood of the X-ray absorbing atom, a single crystal is not required and the method can be successfully used both with small size crystallites and even amorphous solids. In the case of extremely small microcrystals and clusters the mean coordination number around a given atomic species is smaller than that oberved in massive solids. The disadvantage is, that the method requires very intense X-ray sources, but the increasing accessibility of tunable synchrotron radiation in recent years has eliminated the need to use Bremsstrahlung radiation from conventional X-ray tubes, thus drastically shortening the time taken to record a spectrum with consequent improvements in signal-to-noise ratio.

Due to these advantages, a significant amount of work has already been performed on highly dispersed metals with investigations of the local chemical composition, structure and determinations of the size distributions of the smallest clusters. In particular, supported platinum and copper catalysts have been investigated with this technique [46, 47, 48] and interesting data have been obtained on the interatomic distances in small metallic clusters and changes due to the size effect and the influence of adsorbed gases.

REFERENCES

[1] F. Sauerwald and S. Kubik, *Z. Elektrochem.*, **38**, 33 (1932).
[2] W. Trzebiatowski, *Z. Physik. Chem.*, **A169**, 91 (1934).
[3] G. Grube and H. Schlecht, *Z. Elektrochem.*, **44**, 367 (1938).
[4] E. Raub and W. Plate, *Z. Metallkd.*, **40**, 171, 206 (1949).

[5] J. Martinet and S. Tacvorian, *Rev. Métallurgie*, **54**, 65 (1957).
[6] R. M. Hill, *Proc. Roy. Soc.*, **A309**, 377 (1969).
[7] C. A. Neugebauer and M. B. Webb, *J. Appl. Phys.*, **33**, 74 (1962).
[8] T. E. Hartman, *J. Appl. Phys.*, **34**, 943 (1963).
[9] R. M. Hill, *Nature*, **204**, 35 (1964).
[10] R. M. Hill, *Proc. Roy. Soc.*, **A309**, 397 (1969).
[11] J. E. Morris and T. J. Coutts, *Thin Solid Films*, **47**, 81 (1977).
[12] J. Frenkel and J. Dorfman, *Nature*, **126**, 274 (1930).
[13] L. Néel, *Compt. Rend. Acad. Sci.*, **228**, 664 (1949); *Revs. Mod. Phys.*, **25**, 293 (1953).
[14] S. W. Wonsowski, *Magnetizm*, Izd. Nauka, Moscow 1971, p. 803.
[15] P. W. Selwood, *Adsorption and Collective Paramagnetism*, Academic Press, New York–London 1962.
[16] W. Romanowski, *Roczniki Chem.*, **34**, 239 (1960); *Z. anorg. allg. Chem.*, **351**, 180 (1967).
[17] H. Dreyer, *Z. anorg. allg. Chem.*, **362**, 233 (1968).
[18] J. T. Richardson and P. Desai, *J. Catal.*, **42**, 294 (1976).
[19] H. Kubicka, *J. Catal.*, **5**, 39 (1966).
[20] R. Denton, B. Muhlschlegel and D. J. Scalapino, *Phys. Rev. Letters*, **26**, 707 (1971); *Phys. Rev.*, **B7**, 3589 (1973).
[21] R. E. Marzke, W. S. Glaussinger and M. Bayard, *Solid State Comm.*, **18**, 1025 (1976).
[22] R. D. Maurer, *J. Chem. Phys.*, **25**, 1206 (1956).
[23] R. D. Maurer, *J. Appl. Phys.*, **29**, 1 (1958).
[24] M. A. Smithard and R. Dupree, *Phys. Status Sol.*, **11a**, 695 (1972).
[25] J. A. A. J. Perenboom, P. Wyder and F. Meier, *Phys. Rep.*, **78**, 2 (1981).
[26] H. Klug and L. E. Alexander, *X-Ray Diffraction Procedures for Polycrystalline and Amorphous Materials*, sec. ed., J. Wiley, New York 1974, p. 635.
[27] B. E. Warren and B. L. Averbach, *J. Appl. Phys.*, **21**, 595 (1950).
[28] G. K. Williamson and W. Hall, *Acta Metalurg.*, **1**, 22 (1953).
[29] A. Bienenstock, *J. Appl. Phys.*, **32**, 187 (1961); *ibid.*, **34**, 1391 (1963).
[30] A. Guinier, *Théorie et technique de la radiocrystallographie*, Dunod, Paris 1956, p. 636 ff.
[31] A. Guinier, *Compt. Rend. Acad. Sci.*, **204**, 1115 (1937); *Ann. physiq.*, **12**, 161 (1939); *J. chim. physiq.*, **40**, 133 (1943).
[32] G. Porod, *Kolloid Z.*, **124**, 83 (1951); *ibid.*, **125**, 51, 109 (1952); in *Small Angle X-Ray Scattering*, Ed. H. Brumberger, Gordon and Breach, New York–London 1967, p. 1–15.
[33] O. Kratky, *Oesterr. Chem. Ztg.*, **54**, 193 (1953) and references therein.
[34] M. H. Jellinek, E. Solomon and I. Fankuchen, *Ind. Eng. Chem. Anal. Ed.*, **18**, 172 (1946).
[35] S. H. Bauer, *J. Chem. Phys.*, **13**, 450 (1945).
[36] S. D. Harkness, R. Gould and J. J. Hren, *Phil. Mag.*, **19**, 115 (1969).
[37] V. Gerold, in *Small Angle X-Ray Scattering*, *loc. cit.*, p. 292 ff.
[38] G. A. Samorjai, R. E. Powell, P. W. Montgomery and G. Jura, *ibid.*, p. 449 ff.
[39] A. Renouprez and B. Imelik, *J. Appl. Crystallogr.*, **6**, 105 (1973).
[40] A. Renouprez, C. Hoang Van and P. A. Compagnon, *J. Catal.*, **34**, 411 (1974).
[41] P. Gallezot, A. Alarcon-Diaz, J. A. Dalmon, A. Renouprez and B. Imelik, *J. Catal.*, **39**, 334 (1975).

[42] O. Kratky, *Z. Elektrochem.*, **58**, 49 (1954).

[43] R. L. Kronig, *Z. Physik*, **70**, 317 (1931); *ibid.*, **75**, 191 (1932); *ibid.*, **75**, 468 (1932).

[44] D. E. Sayers, F. W. Lyttle and E. A. Stern, *Adv. X-Ray Analysis*, **13**, 248 (1970); *Phys. Rev. Lett.*, **27**, 1204 (1971).

[45] C. A. Ashley and S. Doniach, *Phys. Rev.*, **B11**, 1278 (1975); P. A. Lee and J. B. Pendry, *ibid.*, **B11**, 2795 (1975).

[46] G. H. Via, J. H. Sinfelt and F. W. Lyttle, *J. Chem. Phys.*, **71**, 690 (1979).

[47] S. J. Gurman, *J. Mat. Sci.*, **17**, 1541 (1982).

[48] J. F. Hamilton, G. Apai, S. T. Lee and M. G. Mason, in *Growth and Properties of Metal Clusters*, Ed. J. Bourdon, Elsevier Scientific Lublishing Company, Amsterdam 1980, p. 387.

[49] A. Renouprez, P. Foilloux and B. Moraveck, *ibid.*, p. 421.

4

Geometry and Electronic Structure of Small Metallic Particles

The arrangement of atoms in very small solid particles containing up to a few tens of atoms cannot be seen directly even with the aid of good electron microscopes *. Thus our ideas about the symmetry and geometric properties of such ensembles of identical atoms must, of necessity, be based on analogies and results relating to larger crystallites. It seems natural to assume that the smallest crystalline particles are built up according to the same rules of geometric crystallography which apply to large crystals. Since the majority of metals crystallize in simple, highly symmetrical structures: face centred cubic (A_1), body centred cubic (A_2) and hexagonal close packed (A_3), models of small metallic particles have been constructed in the form of polyhedra containing a given number of atoms and constituting fragments of the corresponding space lattices. The atoms themselves were considered to the rigid spheres, according to the convention adopted in the crystallography of metals.

Studies of the arrangement of hard spheres in space — so important in crystallography — have also played an important part in suggesting candidate structures and in providing estimates of the stability of aggregates containing a small number of atoms. The stability condition for

* Recent improvements in high resolution electron microscopes have made it possible, however, at least in selected cases, to resolve the internal structure even of very small particles (1–2 nm in size). A recent review of the relevant experimental work is contained in Vol. 20 of the *Journal of Molecular Catalysis* (September 1983).

such aggregates is simply: for a given number of atoms, the potential energy — or, more generally, the free energy — shall be a minimum. This kind of approach does not, of course, exclude consideration of polyhedra other than the "lattice" polyhedra of classical crystallography and, in fact, it includes even disordered (amorphous) aggregates.

Using a realistic potential function, a minimum of the potential energy would be expected for the most dense by packed arrangement of atoms — considered as hard spheres. Since it is known that the densest packing of spheres in a translationally periodic solid is obtained in the A_1 and A_3 structures, polyhedra constituting fragments of these structures were the first to be considered.

4.1 "LATTICE" STRUCTURES OF SMALL ATOMIC AGGREGATES

From a given small number of atoms, simple, regular polyhedra can be formed, which fill space efficiently and have at least some of the symmetry elements of the most densely packed lattices. Such polyhedra for the A_1 lattice are the tetrahedron, octahedron, cube and cuboctahedron. The minimum number of atoms needed to construct these polyhedra, are four, six, fourteen and thirteen, respectively (Fig. 4.1), but clearly each of these polyhedra can be constructed using a larger number of atoms. Similarly, the corresponding polyhedra for the A_2 structure are: cube (9 atoms), tetragonal bipyramid (6 atoms) and rhombic dodecahedron (15 atoms). For the A_3 structure the simplest lattice polyhedra are the hexagonal prism, hexagonal bipyramid, and truncated hexagonal bipyramid. A list of the numbers of atoms included in such polyhedra of increasing dimensions is given in one of the earlier works of this author [1].

Other authors [2, 3] have considered similar polyhedra in connection with the investigation of the properties of adsorption sites in catalysts, consisting of very small metallic crystallites.

In the first of these papers, a minimum in the surface energy has been assumed as a criterion for the stability of polyhedra. Using this constraint, an energy minimum occurs in the A_1 structure for a cuboctahedron, in the A_2 structure for the rhombic dodecahedron, and in the A_3 structure for the truncated hexagonal bipyramid. In the following sections, we see that these results have only a limited value, due to the approximations involved in the hard sphere model and the neglect of other possible densely packed polyhedra with symmetries other than those found in the A_1, A_2 and A_3 structures.

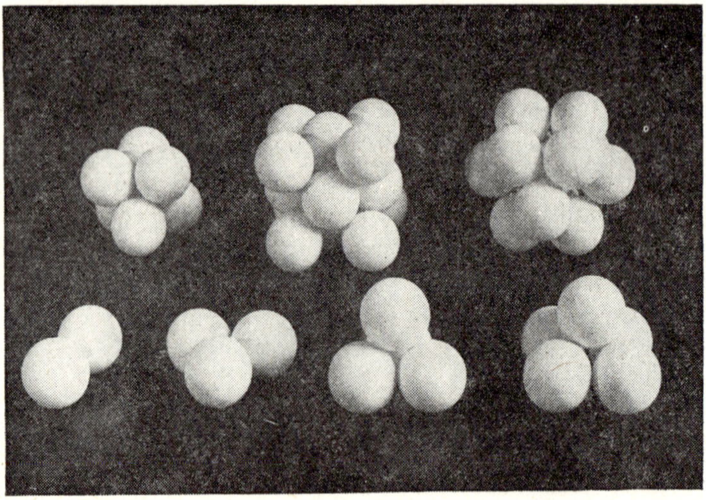

(a)

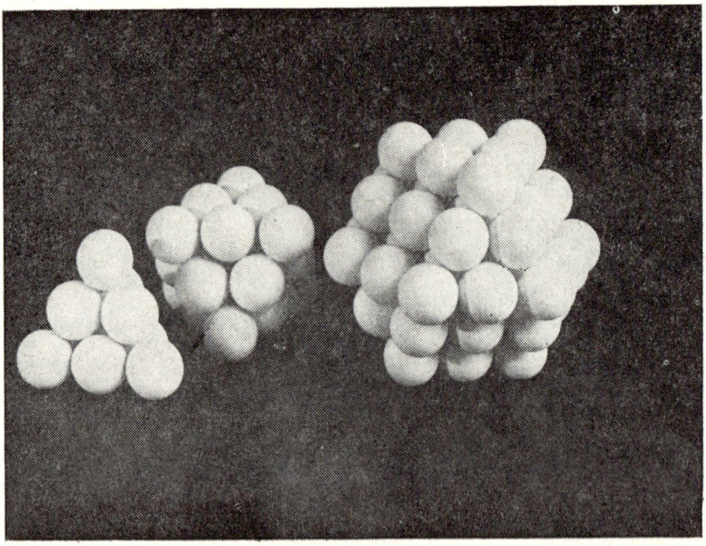

(b)

Fig. 4.1 — Small clusters observed in A_1 lattice structures: (a) bottom row: pair ($n = 2$), equilateral triangle ($n = 3$), tetrahedron ($n = 4$), tetragonal pyramid ($n = 5$); top row: octahedron ($n = 6$), cube ($n = 14$), cuboctahedron ($n = 13$); (b) tetrahedron ($n = 10$), octahedron ($n = 19$), cuboctahedron ($n = 55$).

4.2. THE STABILITY OF SMALL ATOMIC AGGREGATES

4.2.1 Potential energy minimum

For "clusters" containing up to a few tens of atoms, it is possible to calculate the minimum potential energy or free energy using fairly exact, commonly accepted computational procedures. The geometrical configuration is varied and the configuration, which has the smallest potential energy (or free energy) for the potential energy function chosen is considered to be the most stable.

The problem of the interaction between many atoms, known as the many body problem, becomes calculable with an assumption of some simplified model of interaction. A commonly adopted approximation is to express the total interaction as a sum of interactions between all possible pairs of atoms (pairwise interaction). There are $n(n-1)/2$ pairs, where n is the number of interacting bodies — in our case the number of atoms in a cluster. Neglecting the mutual interaction of all other atoms with a given pair, the potential energy becomes

$$E(r_1, r_2, \ldots, r_{n(n-1)}) = \sum_i \sum_{j>i} V(|r_i - r_j|), \tag{4.1}$$

where $V(|r_i - r_j|)$ is an interaction potential, dependent on the distance vectors r_{ij} between atoms. Usually, one of the following analytical forms of potential is assumed:

The *Morse potential*

$$V(r) = \{1 - \exp[a(1-r)]\}^2 - 1, \tag{4.2}$$

or the *Mie potential*

$$V(r) = \frac{1}{m-n}(nr^{-m} - mr^{-n}), \tag{4.3}$$

where r is the distance between a pair of atoms and a, m and n are constants (usually integers). The Mie potential is more frequently used, especially with $n = 6$ and $m = 12$ (Lennard–Jones potential). In equations (4.2) and (4.3) the r-value are scaled by division with r_0, the interatomic distance at equilibrium, so that for $r = r_0$, $V(r) = -1$.

Calculations of the potential energy for ordered atom aggregates using these potential functions have been carried out frequently; perhaps most complete being those of Allpress and Sanders [4] and Hoare and Pal [5]. The choice of the geometric configurations corresponding to potential energy minima presents no difficulty in the case of clusters of 3 and 4 atoms, for which the configurations are, clearly, the equilateral triangle and the tetrahedron, respectively. However, beginning with $n = 5$ there are a number of possible "isomers" — that is, ordered configurations, at local minima of potential energy. The number of isomers increases rapidly with increasing n, so that until recently, no calculations were possible for $n > 100$. In fact, there is no certainty that the minima found for $n < 100$ represent absolute minima. It is certain however, that relative minima can be found. These are definite configurations for which the potential energy is smaller than for others tested. Such configurations are frequently essentially different from the "lattice" polyhedra typical of the densely packed lattices, mentioned in the preceding section. They can be considered to be related to the latter by "relaxation", that is, by small changes of interatomic distances relative to the fixed distances of structures composed of hard spheres.

Allpress and Sanders [4] have shown that the binding energy per atom is larger and the number of surface atoms lower for icosahedra containing up to 2000 atoms than for corresponding octahedra containing the same number of atoms, regardless of the values of the constants assumed in the potential functions (4.2) and (4.3). Thus an icosahedral cluster with fivefold symmetry axes appears to be the most stable. Fivefold symmetry is not encountered in hard sphere lattices; its existence is only possible as a result of the distortion of interatomic distances allowed by a "soft" potential energy function.

Hoare and Pal [5] have established that energetically optimal clusters are obtained for $n \leqslant 55$, which are not the lattice polyhedra of the A_1 and A_3 structures but can be obtained by multiple twinning of tetrahedra and relaxation of the consequent strain. In particular, the cuboctahedra with $n = 13$ and $n = 55$, proposed by Romanowski and others as optimal lattice polyhedra have been shown to form icosahedra on relaxation, with larger binding energies and slightly smaller volumes [6].

Polyhedra with fivefold symmetry axis composed of densely packed spheres were also considered by Bagley [7] and by Fukano and Wayman [8], who tried to demonstrate the extraordinary stability of icosahedral clusters. The structures of densely packed spheres having fivefold symmetry axes, discussed by Bagley have a packing fraction equal to 0.7236, only slightly smaller than the value for the most densely packed A_1 and A_3 classical structures (0.7405). The occurrence of particles with fivefold symmetry such as icosahedra was supported by the kinetic arguments of Gillet [9] who demonstrated that the growth of such structures initiated on any plane of an icosahedral nucleus is much faster than the growth of A_1 type lattice structures under the same conditions. Moreover, Gillet showed that icosahedral clusters are stable to dimensions of 8–10 nm, and that only clusters exceeding these dimensions transform into the A_1 structure. Models for some clusters with fivefold symmetry axes are shown in Fig. 4.2.

4.2.2 Thermodynamic criteria for cluster stability

Minimization of the potential energy represents a satisfactory criterion for cluster stability only in the absence of kinetic energy — at absolute zero. Therefore at finite temperatures, it is more correct to postulate a minimum of free energy which takes into account the entropy and temperature factors. The entropy and free energy of systems composed of small numbers of atoms can be calculated by elementary statistical thermodynamics via the partition function, Z, of the system composed of the respective partition functions for translation, rotation and vibration of the system

$$Z(T) = Z_{tr} Z_{vibr} Z_{rot} \exp\left(-\frac{E_0}{kT}\right). \tag{4.4}$$

Here, Z_{tr}, Z_{vibr} and Z_{rot} are the partial partition functions for translation, vibration and rotation respectively, and E_0 is the binding energy at the equilibrium position around which vibration occurs. Assuming a harmonic oscillator approximation for vibration and rigid body rotation then

$$Z_{tr} = \left(\frac{2mnkT}{h^2}\right)^{3/2} V, \tag{4.5}$$

$$Z_{rot} = \frac{\pi^{1/2}}{\sigma}\left(\frac{8\pi kT}{h^2}\right)^{3/2} (I_A I_B I_C)^{1/2}, \tag{4.6}$$

$$Z_{vib} = \prod_{i=1}^{3n-6} \exp\left(-\frac{h\nu_i}{2kT}\right)\left[1 - \exp\left(-\frac{h\nu_i}{kT}\right)\right]^{-1}. \tag{4.7}$$

Here, m is the mass of a single atom, V is the effective volume of the system, ν_i are normal vibration frequencies around the equilibrium position, σ is a number representing the rotational symmetry of the cluster (equal to 24 for a cubic symmetry), I_A, I_B and I_C are the principal moments of inertia of the cluster and n is the number of atoms.

Thermodynamic functions and the partition functions are related as follows:

Energy

$$E = kT^2 \frac{\partial \ln Z}{\partial T}, \tag{4.8}$$

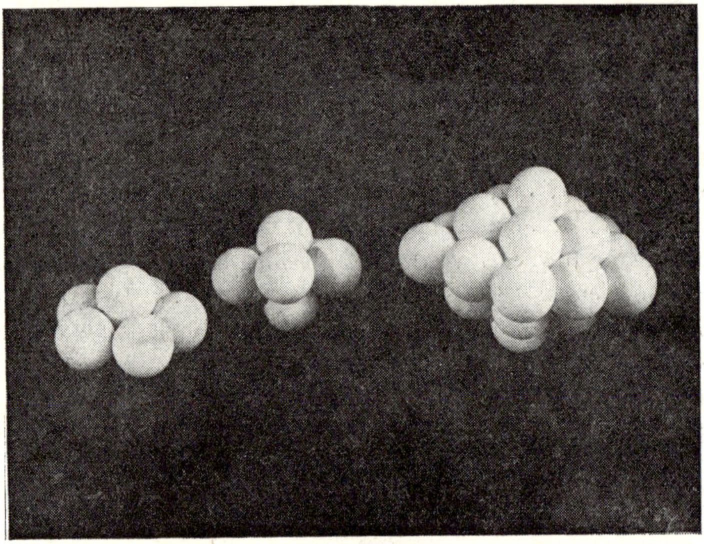

(a)

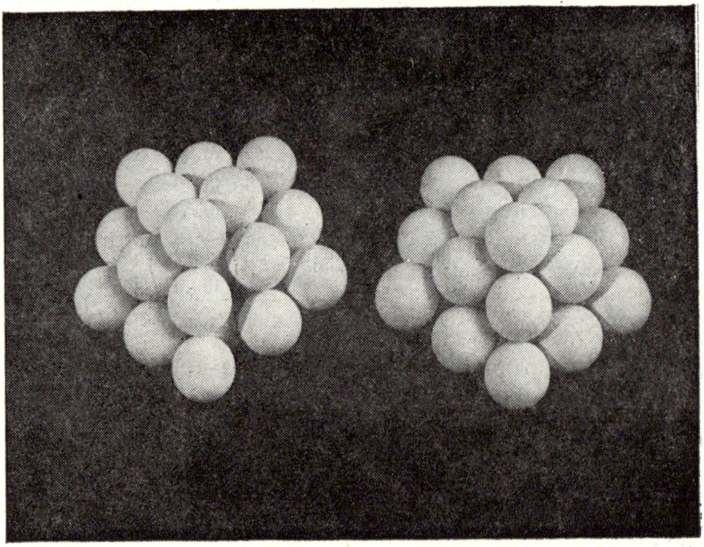

(b)

(c)

(d)

Fig. 4.2 — Clusters with pentagonal symmetry: (a) pentagonal pyramid ($n = 6$), pentagonal bipyramid ($n = 7$) and pentagonal bipyramid ($n = 23$): (b) left: pentagonal bipyramid obtained by twinning of 5 tetrahedra, right: the same after "relaxation"; (c) cuboctahedron ($n = 13$) and corresponding icosahedron; (d) cuboctahedron ($n = 55$) and corresponding icosahedron.

Entropy

$$S = kT\frac{\partial \ln Z}{\partial T} + k\ln Z\,, \tag{4.9}$$

Free energy

$$F = -kT\ln Z\,, \tag{4.10}$$

Heat capacity

$$C_p = 2kT\frac{\partial \ln Z}{\partial T} + kT^2\frac{\partial^2 \ln Z}{\partial T^2}\,. \tag{4.11}$$

For a metallic cluster deposited on a substrate and interacting with the latter with a given force, we can neglect the translational and rotational motion of the cluster as a whole, but it is necessary to take into account the additional degrees of freedom, associated with the vibration of the entire cluster relative to the substrate. In the case of clusters containing more than a few atoms, the energy per atom of these vibrations is small and can be neglected, so that the general partition function will be purely vibrational

$$Z(T) = \exp\left(-\frac{E_0}{kT}\right)\prod_{i=1}^{3n-6}\frac{\exp(-h\nu_i/2kT)}{1-\exp(-h\nu_i/kT)}\,. \tag{4.12}$$

The free energy of vibration is then

$$F_{\mathrm{vib}} = E_0 + \varepsilon_0 + kT\sum_{i=1}^{3n-6}\ln\left[1-\exp\left(-\frac{h\nu_i}{kT}\right)\right], \tag{4.13}$$

where ε_0 is the zero point energy for $3n-6$ vibrations

$$\varepsilon_0 = \frac{1}{2}\sum_{i=1}^{3n-6} h\nu_i\,. \tag{4.14}$$

Thus, by comparing the calculated vibrational free energies for various isomeric clusters — that is, clusters with the same number of atoms, but with different geometries — we can choose those with the smallest free energy and consequently, which represent the most stable configuration.

Such calculations have been carried out recently by a number of investigators [5, 6, 10–13]. Clusters in the form of lattice polyhedra of the most densely packed structures have been considered, as well as non-crystallographic and spherical clusters obtained by the relaxation of the lattice structures. For the case of clusters containing less than 100 atoms, for which calculations are possible with current computers, it has been established that a minimum of free energy occurs for icosahedral clusters with $n = 13$ and $n = 55$. Also, calculations of the heat capacity and entropy reaffirmed the particular stability of these clusters [5, 14].

4.3. EXPERIMENTAL PROOF OF THE EXISTENCE OF METAL PARTICLES WITH AND WITHOUT LATTICE SYMMETRIES

Theoretical determination of the geometry of clusters corresponding to minima in the potential energy, or free energy, are possible only for a small number of atoms, so that the cluster size lies at the lower limit of resolution of even the best electron microscopes. Therefore the results of the theoretical calculations mentioned above, can rarely be directly verified experimentally by this technique. Historically however, information on the non-crystallographic structure of certain metallic particles was first obtained experimentally, and it was only afterwards that theoretical approaches and calculations demonstrated the stability of such particles.

Fivefold symmetry in cobalt crystals, obtained by hydrogen reduction of cobalt bromide has been observed even by optical microscopy [15], since the crystals were up to 30 μm in size. Fivefold symmetry has also been observed in dendritic whiskers of copper, obtained by electrolysis [16], and using the *field electron emission method* (FEM), 5-fold axes have been found in whiskers of iron, nickel and platinum, grown on various metallic substrates [17].

High resolution electron microscope images of crystals in the form of pentagonal pyramids and icosahedra have been observed in the early stages of formation of thin metal films evaporated in vacuum [18, 19]. Kimoto and Nishida [20] have observed these polyhedra with dimensions up to 100 nm in gold and silver films evaporated in an argon atmosphere at low pressures. Elastic strain at the twin boundaries of tetrahedra has been observed in pentagonal bipyramids by electron microscopy [21]. Formation of pentagonal bipyramids by the twinning of tetrahedral nuclei had been suggested earlier by Ino [18]. Fukano and Wayman [8] observed however, that formation does not necessarily proceed by twinning of tetrahedral nuclei, since the simplest pyramid composed of six atoms (see Fig. 4.2) can also act as a nucleus for crystallization.

In their extensive electron microscopic studies of small palladium and platinum particles evaporated in ultra-high vacuum onto a cleaved NaCl substrate, Gillet and Renou [22] observed that particles having dimensions in the range of 6–20 nm had cubic and rhombohedral shapes. The external planes were (sometimes incomplete) {111} --planes of the A_1 structure. Particles with dimensions below 2 nm were single crystals with the fcc structure, with shapes which could not be exactly determined, due to their small dimensions. In earlier work, Gillet established [23] however, that silver and gold particles of approximately the same size showed a pentagonal symmetry and an icosahedral structure. For sizes greater than 8 nm, icosahedral particles changed into multiply twinned structures with a pseudopentagonal structure. These are composed of fcc tetrahedra with strain at twinning boundaries.

In a series of studies using electron microscopy, carried out by Uyeda and co-workers [24, 25] on metal microparticles formed by thermal evaporation and condensation in an inert gas atmosphere (see Chapter 2) it was established that A_1-metals like cobalt, nickel, copper, palladium, silver and gold form distinct shapes of which the truncated octahedral and cuboctahedral particles were true single crystals, while icosahedral and pentagonal bipyramids were composed of multiply-twinned tetrahedra. However, the particles obtained by Uyeda *et al.* were comparatively large (> 50 nm) that is, larger than those obtained by vacuum evaporation.

For small particles of metals with the bcc structure: vanadium, chromium, iron and molybdenum, the principal polyhedron was observed the rhombic dodecahedron with a variable degree of truncation, characteristic of the A_2 structure. Other shapes (rounded cube, icositetrahedron) seemed to be associated with the $A15$ structure found in chromium and molybdenum at higher temperatures. During quenching of the high temperature forms, the external shape was preserved, although the structure did not always change into the A_2 modification [26, 27]. Formation of this untypical $A15$ structure seems to occur for smaller Mo and Cr crystals only; the bigger ones convert easily during quenching into normal A_2 structures [25].

From the experimental data presented above, it is clear that smal particles of metals occur both as crystallographic (Wulff polyhedra) and non-crystallographic polyhedra. The non-crystallographic polyhedra seem to be preferred as equilibrium forms for the smallest particles, while for larger particles the external shape corresponding to non-crystallographic symmetry is preserved, the internal structure being that of regular structures however.

In connection with the problem of crystallographic and non-crystallographic equilibrium forms it should be noted that the smallest particles do not always form in conditions which allow them to reach true equilibrium. In most cases nucleation and growth proceed at high supersaturation and on substrates which can induce the structure of crystallization nuclei. In addition, the differences in energy between crystallographic and non-crystallographic structures may be too small to bring about the structural transition associated with absolute equilibrium. Rapid quenching of small crystals, which occurs in most formation processes, does also not favour attainment of true equilibrium.

4.4 ELECTRONIC STRUCTURE OF THE SMALLEST METAL PARTICLES

A solid metallic phase and a single metal atom have different electronic structures, which are described by allowed energy bands for crystals and discrete energy levels for single atoms. One could expect, therefore, that very small metal particles, composed of a small number of atoms would have an electronic structure intermediate between the two limiting cases.

In a macroscopic metal crystal, individual energy levels within an allowed energy band are very closely spaced with a separation δ, proportional to l^{-3}, where l is a linear dimension of the crystal. For small crystals Kubo [28], has shown that δ is approximately given by the relation

$$\delta = 4\varepsilon_F/3N, \tag{4.15}$$

where ε_F is the Fermi energy and N is the number of atoms in the crystal. For $\varepsilon_F \simeq 10$ eV and $N \approx 10^3$, $\delta \approx 1.4 \times 10^{-2}$ eV, a quantity which is of the same order of magnitude as the thermal energy kT at room temperature. For $N = 10^5$ the value of δ is of the order of 10^{-4} eV, approximately equal to the thermal energy at 1 K. Electron energy level spacings of this magnitude suggest the occurrence of some anomalies in the electronic properties of small particles compared with macroscopic crystals. For the smallest particles, and assuming the validity of equation (4.15), still larger electron level spacings, would be expected, approaching those found in single atoms. It is clear however, that such an extrapolation could prove to be risky and that the calculation of electron energy levels in clusters consisting of few atoms should be approached in a more rigorous way.

4.4.1 Application of the methods of quantum chemistry to calculations of the electronic structure of metallic clusters

Calculation of the electron energy level spacings in metal atom aggregates is possible with the aid of semi-empirical methods using Molecular Orbital (MO) treatments already established in calculations for polyatomic molecules in organic chemistry. These methods were applied for the first time to metallic clusters by R. C. Baetzold and collaborators from Eastman Kodak Laboratory. They compared the results obtained with Extended Hückel Molecular Orbital (EHMO) and Complete Neglect of Differential Overlap (CNDO) methods (see Appendix) with experimentally established properties of massive metals and discussed the differences appearing in these properties between macrocrystals and small clusters.

Properties directly related to the energy of electronic states such as the ionization potential (*IP*) and electron affinity (*EA*) for single atoms, and the corresponding work function for the crystalline metallic phase, are especially suitable for such comparisons. In metal clusters, a quantity which corresponds to the ionization potential is the energy of the highest occupied electron states (Highest Occupied Molecular Orbital — HOMO) and the energy of the Lowest Unoccupied Molecular Orbital (LUMO) corresponds to the electron affinity. Differences between macroscopic metals and small clusters are evident also in the binding energy (*BE*) per atom.

It is clear that owing to the very small distance between electron levels in bulk metals, both ionization potential and electron affinity become equivalent and equal to the work function, φ

$$IP = EA = \varphi \,. \tag{4.16}$$

For a single atom, where the spacing between the highest occupied and lowest unoccupied level is large,

$$EA < IP \,. \tag{4.17}$$

The binding energy per atom can be expressed as

$$\frac{BE}{n} = -\frac{E_c}{n} + E_a \,, \tag{4.18}$$

where E_c stands for the total energy of a cluster and E_a is the energy of a single atom.

Baetzold carried out a series of EHMO calculations of the electron energy levels of transition metal clusters as well as of those of silver and cadmium, containing firstly up to 30 atoms [29–31] and subsequently up to 55 atoms [32]. He determined the positions of the LUMO (equivalent to the electron affinity), the HOMO (ionization level) and the binding energy per atom in these clusters. The final results of these calculations may be summarized as follows:

1. The binding energy per atom for silver clusters is higher for linear and two-dimensional aggregates than for three-dimensional clusters for $n \leqslant 55$ indicating the greater stability of linear aggregates (Fig. 4.3).

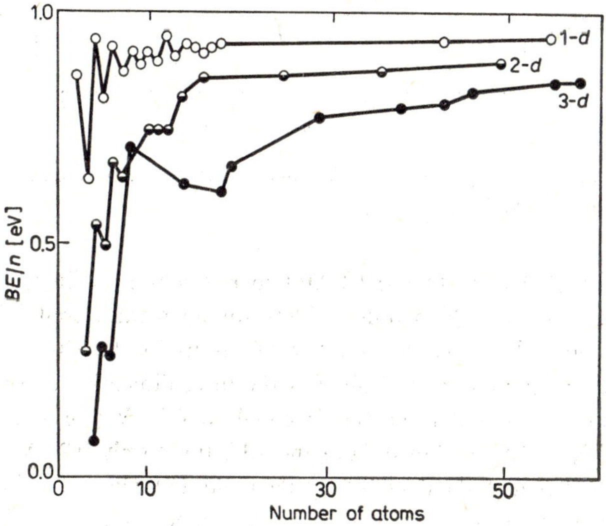

Fig. 4.3 — Binding energy per atom for linear (1-*d*), two-dimensional (2-*d*) and three-dimensional Ag-clusters, calculated by the Extended Hückel Molecular Orbital method (after N. Saito *et al.* [26]).

This result is somewhat surprising and was even not given full credit by the author himself. It was obtained using distances between silver atoms equal to 0.265 nm for the linear aggregate and 0.325 nm for three-dimensional clusters. These distances differ significantly from those found in bulk silver (A_1 lattice) (0.288 nm) and were assumed to be compatible with the minimization of energy for the given geometric configuration. This author is of the opinion that the Extended Hückel method is not suitable for the determination of the equilibrium positions of atoms.

The same situation prevails for palladium clusters with a small number of atom. However, above $n \approx 10$, the binding energy in the three-dimensional cluster is higher than in the linear aggregate (Fig. 4.4).

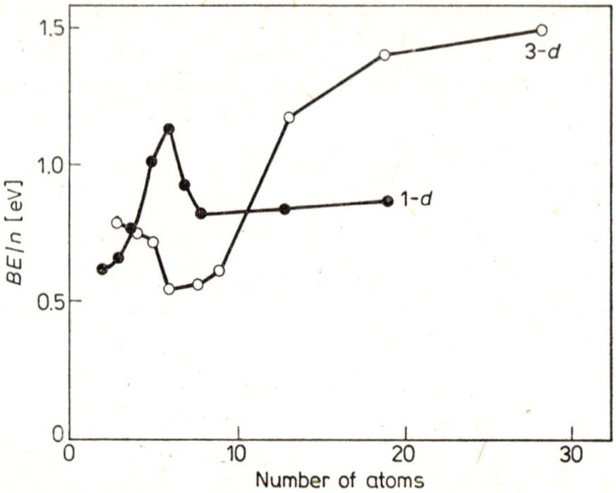

Fig. 4.4 — Binding energy per atom for linear (1-*d*) and three-dimensional clusters of palladium (after N. Saito *et al.* [26]).

2. Calculated HOMO and LUMO energy values for linear and three-dimensional models, which differ widely for aggregates containing a few atoms, become closer as the number of atoms in the cluster increases, as if approaching a common limit. For the three-dimensional Ag_{55} cluster, the ionization potential (HOMO) is equal to 6.1 eV (for massive silver $\varphi = 4.3$ eV), and its distance from the LUMO is only 0.2 eV. Values of the ionization potential calculated by Baetzold and Mack [32] for 19-atom crystallographic clusters of silver, gold, copper and palladium differ significantly from the corresponding work functions of these metals, as do the energies of the *d*-bands and their widths (Fig. 4.5).

Baetzold has also applied the CNDO method for the calculation of the electronic properties of transition metal clusters [30] and of silver [33], deposited on substrates of carbon and SiO_2 [34]. In these calculations, a number of substrate atoms were included in the quantum mechanical system, consequently, an estimate of the interaction between the cluster and substrate was possible. Calculations for "alloy clusters" containing atoms of two metals have also been carried out for Ag-Pd [32], Ni-Cu [31, 32, 35] and Ni-Pd [31].

Differences between the interatomic distances and binding energies calculated by Baetzold and other authors, and the corresponding values observed in larger crystals are significant. They are due, at least in part, to the approximations involved in semi-empirical methods and to the choice of the arbitrary parameters applied therein.

Blyholder [36] attempted to improve this situation by choosing the arbitrary parameters in such a way that the interatomic equilibrium distance, the population of the *d*-band, the binding energy and the ionization potential for an octahedral Ni_6 cluster were close to the corresponding values for bulk nickel. The results, obtained using these assumptions for cluster containing up to 13 atoms always predicted a greater stability for three-dimensional clusters, as compared to one- and two-dimensional clusters. Moreover, these calculations, carried out by the CNDO method seem to confirm the rule usually assumed without discussion [1], that the binding energy per atom is proportional to the number of nearest neighbours (coordination number).

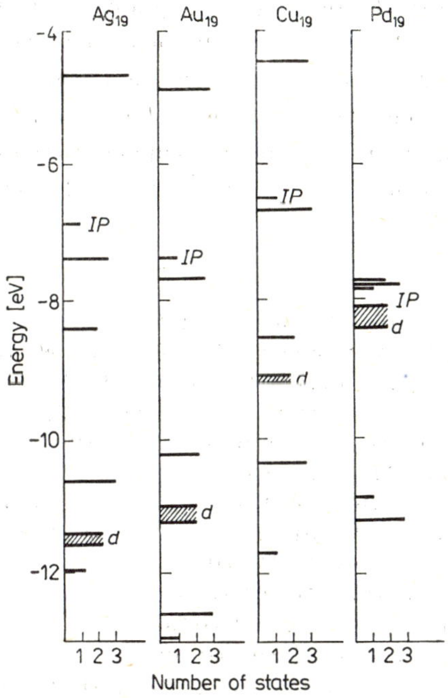

Fig. 4.5 — Ionization potential and *d*-band positions for 19-atom clusters of the group IB metals and palladium (after N. Saito *et al.* [26]).

There are a number of other methods for calculating the electronic structure of small metallic clusters, which are considered to be more rigorous than EH or CNDO methods and yet are computationally manageable. On of them is the so-called SCF-X_α-SW (Self-Consistent Field X_α-Scattered Wave) method developed by Slater and Johnson [37] (see

Appendix). This method has been used to calculate the electronic levels of 13-atom cuboctahedral and icosahedral clusters of the A_1 metals [38]. A so-called Self-Consistent Hartree–Fock–Slater (SCHFS) method has been applied by Tanabe *et al.* [39] for the calculation of the electronic levels of Ni_5, Ni_9, Pd_9 and Pd_{19} A_1 lattice clusters. The level structures of these clusters were given as density of states curves, which are similar to ultraviolet photoemission spectra (see Chapter 6) even for 9-atom clusters. The larger octahedral Pd_{19} cluster was found to have certain bulk properties comparable to the bulk metal and also to be in good agreement with UPS spectra. Hamilton and Baetzold [40] in their review of results obtained by EH and CNDO methods assert that the change from the set of quasi-atomic electron levels to the characteristic "density of states" spectrum of the bulk metal is gradual and that the spectrum of a cluster of more than 100 atoms becomes indiscernible from that of the bulk metal. In the special case of palladium, they observed that the distribution of electrons between *d* and *s* states is close to that bulk palladium even for clusters of 8–10 atoms, changing from a $4d^{10}$ configuration in the single atom to $4d^{9.4}s^{0.6}$, characteristic of the bulk metal.

REFERENCES

[1] W. Romanowski, *Surface Sci.*, **18**, 373 (1969).
[2] O. M. Połtorak and W. S. Boronin, *Zh. Fiz. Khim.*, **40**, 2671 (1966).
[3] R. van Hardeveld and F. Hartog, *Surface Sci.*, **15**, 189 (1969).
[4] J. G. Allpress and J. V. Sanders, *Austral. J. Phys.*, **23**, 23 (1970).
[5] M. R. Hoare and P. Pal, *J. Cryst. Growth*, **17**, 77 (1972).
[6] J. J. Burton, *Nature*, **229**, 335 (1971).
[7] B. G. Bagley, *Nature*, **208**, 674 (1965); *J. Cryst. Growth*, **6**, 323 (1970).
[8] Y. Fukano and C. M. Wayman, *J. Appl. Phys.*, **40**, 1656 (1969).
[9] M. Gillet, *J. Cryst. Growth*, **36**, 239 (1976).
[10] S. G. Reed, *J. Chem. Phys.*, **20**, 208 (1952).
[11] J. J. Burton, *J. Chem. Phys.*, **52**, 345 (1970).
[12] D. J. McGinty, *J. Chem. Phys.*, **55**, 560 (1971).
[13] K. Nishioka, R. Shawyer, A. Bienenstock and G. H. Pound, *J. Chem. Phys.*, **55**, 5082 (1971).
[14] J. J. Burton, *J. Chem. Phys.*, **56**, 3133 (1972).
[15] M. A. Gedwill, C. J. Allstatter and C. M. Wayman, *J. Appl. Phys.*, **35**, 2266 (1964).
[16] F. Ogburn, B. Paretzkin and H. S. Peiser, *Acta Cryst.*, **17**, 774 (1964).
[17] A. J. Melmed and O. D. Hayward, *J. Chem. Phys.*, **31**, 545 (1959).
[18] S. Ino, *J. Phys. Soc. Japan*, **21**, 346 (1966).
[19] J. G. Allpress and T. V. Sanders, *Surface Sci.*, **7**, 1 (1967).
[20] K. Kimoto and I. Nishida, *J. Phys. Soc. Japan*, **32**, 940 (1967).
[21] T. Komoda, *Jap. J. Appl. Phys.*, **7**, 27 (1968).
[22] M. Gillet and A. Renou, *Surface Sci.*, **90**, 91 (1979).

[23] M. Gillet, *Surface Sci.*, **67**, 139 (1977).
[24] R. Uyeda, *J. Cryst. Growth*, **24/25**, 69 (1974)
[25] R. Uyeda, *J. Cryst. Growth*, **45**, 485 (1978).
[26] N. Saito, K. Mihama and R. Uyeda, *Jap. J. Phys.*, **19**, 1603 (1980).
[27] N. Saito, S. Yatsuya, K. Mihama and R. Uyeda, *J. Cryst. Growth*, **45**, 501 (1978).
[28] R. Kubo, *J. Phys. Soc. Japan*, **17**, 975 (1962).
[29] R. C. Baetzold, *J. Chem. Phys.*, **55**, 4363 (1971).
[30] R. C. Baetzold, *Surface Sci.*, **36**, 123 (1972); *J. Solid. State Chem.*, **6**, 352 (1973).
[31] R. C. Baetzold, *J. Solid. State Chem.*, **29**, 129 (1973).
[32] R. C. Baetzold and R. E. Mack, *J. Chem. Phys.*, **62**, 1513 (1975).
[33] M. G. Mason and R. C. Baetzold, *J. Chem. Phys.*, **64**, 271 (1976).
[34] R. C. Baetzold, *J. Phys. Chem.*, **80**, 1504 (1976).
[35] H. Itoh, *J. Phys.*, **F4**, 1930 (1974).
[36] G. Blyholder, *Surface Sci.*, **42**, 249 (1974).
[37] J. C. Slater and K. H. Johnson, *Phys. Rev.*, **B5**, 844 (1972).
[38] K. H. Johnson and R. P. Mesmer, *J. Vac. Sci. Technol.*, **11**, 236 (1974).
[39] T. Tanabe, H. Adachi and S. Imoto, *Jap. J. Appl. Phys.*, **16**, 1097 (1977).
[40] J. F. Hamilton and R. C. Baetzold, *Science*, **205**, 1213 (1979).

Appendix to Chapter 4

(by H. Chojnacki)

A Survey of Selected Quantum Chemistry Methods

Most of the methods of quantum chemistry applied at present are based on the molecular orbital model, known as the LCAO-MO (Linear Combination of Atomic Orbitals — Molecular Orbitals). In this theory it is assumed that the motion of electrons can be described by wave functions common to the entire molecule, which are therefore called *molecular orbitals*. This model thus constitutes a natural generalization of the notion of orbitals used in the theory of the electronic structure of the atom, to the case of molecules.

The Born–Oppenheimer approximation makes it possible to separate the motions of electrons and nuclei in the molecule, so that it becomes possible to examine the behaviour of electrons at fixed distances between the nuclei. The non-relativistic spinless Hamiltonian of the system composed of A nuclei and N electrons can be described in such an approximation by the equation

$$H = -\frac{\hbar^2}{2m}\sum_{\mu=1}^{N}\nabla_{\mu}^2 - \sum_{\alpha=1}^{A}\sum_{\mu=1}^{N}\frac{Z_{\alpha}e^2}{r_{\alpha\mu}} + \sum\sum_{\mu<\nu}\frac{e^2}{r_{\mu\nu}}, \tag{A 4.1}$$

in which the first term denotes the sum of the kinetic energy operators of the electrons, the second is a sum of operators for interactions of electrons μ with nuclei α, and the last represents the sum of operators for Coulomb interactions between electrons with indices μ and ν.

Since any solution of the Schrödinger equation with the operator given in (A4.1) is not possible, investigation of the electronic structure of molecules requires some approximations. One such approximation is the Hartree–Fock–Roothaan model [1], in which the motion of each electron is considered in the field of the nuclei and in the averaged field of the other electrons. For such a situation, the Fock operator can be written in the form

$$H(\mu) = H^{\text{core}} + \sum_{p}^{\text{occ}} (2J_p - K_p), \tag{A4.2}$$

where the term H^{core} is the atomic core operator and the symbol $\sum_{p}^{\text{occ}}$ denotes summation over all occupied molecular orbitals. In the LCAO-MO theory mentioned above, molecular wave functions are approximated by a linear combination of atomic orbitals

$$\psi_p = \sum_j c_{jp} \chi_j, \tag{A4.3}$$

the set of which, $\{\chi_1, \chi_2, \ldots, \chi_n\}$, is a basis set for the method under consideration. Additionally, in equation (A4.2) the Coulomb operator, J_p, and the exchange operator, K_p, are defined by the relations

$$[J_p(\mu)]\psi_q(\mu) = \left[\int \psi_p^*(\nu) \frac{e^2}{r_{\mu\nu}} \psi_p(\nu)\, \mathrm{d}V_\nu\right] \psi_q(\mu), \tag{A4.4}$$

$$[K_p(\mu)\psi_q(\mu)] = \left[\int \psi_p^*(\nu) \frac{e^2}{r_{\mu\nu}} \psi_q(\nu)\, \mathrm{d}V_\nu\right] \psi_p(\mu), \tag{A4.5}$$

where ψ_p and ψ_q are the molecular orbitals defined by (A4.3). Using the Schrödinger equation

$$H\psi_p = E_p \psi_p \tag{A4.6}$$

and applying the energy operator (A4.2) together with the linear combination (A4.3), we obtain, for the case of a closed-shell system, the relations

$$\sum_j c_{jp} H_{ij} = E_p \sum_j c_{jp} S_{ij} \tag{A4.7}$$

known as the *Hartree–Fock–Roothaan equations* [1]. In these equations

$$S_{ij} = \int \chi_i^* \chi_j \, dV \tag{A4.8}$$

denotes the overlap integral, while

$$H_{ij} = H_{ij}^{\text{core}} + \sum_k \sum_l P_{kl}[(ij|kl) - \tfrac{1}{2}(ik|jl)] \tag{A4.9}$$

are the *matrix elements of Hartree–Fock–Roothaan*. The integral H_{ij}^{core} is defined by relation

$$H_{ij}^{\text{core}} = \int \chi_i^* H^{\text{core}} \chi_j \, dV \tag{A4.10}$$

and

$$P_{kl} = 2 \sum_p^{\text{occ}} c_{kp} c_{lp} \tag{A4.11}$$

is the element of the reduced first-order density matrix. Two-electron repulsion integrals $(ij|kl)$ and $(ik|jl)$ equal to

$$(ij|kl) = \iint \chi_i^*(\mu) \chi_j(\mu) \frac{e^2}{r_{\mu\nu}} \chi_k^*(\nu) \chi_l(\nu) \, dV_\mu dV_\nu , \tag{A4.12}$$

$$(ik|jl) = \iint \chi_i^*(\mu) \chi_k(\mu) \frac{e^2}{r_{\mu\nu}} \chi_j^*(\nu) \chi_l(\nu) \, dV_\mu dV_\nu \tag{A4.13}$$

define the Coulomb integral and exchange integral, respectively.

The total wave function of the N-electron system (normalized and antisymmetric) for the given electron configuration A of the molecule is constructed in the form of a Slater determinant, whose elements are spin orbitals, formed of occupied molecular orbitals and corresponding spin contributions

$$\Psi_A = \frac{1}{\sqrt{N!}} \begin{vmatrix} (\psi_1\alpha)^1 & (\psi_1\beta)^1 & (\psi_2\alpha)^1 & \dots \\ (\psi_1\alpha)^2 & (\psi_1\beta)^2 & (\psi_2\alpha)^2 & \dots \\ (\psi_1\alpha)^3 & (\psi_1\beta)^3 & (\psi_2\alpha)^3 & \dots \\ \dots & \dots & \dots & \dots \end{vmatrix} \tag{A4.14}$$

According to the definition of the expectation value of the energy, the energies of separate electronic levels of the molecule, E_p, corresponding to molecular orbitals ψ_p in (A4.14) are equal to

$$E_p = \int \psi_p^* H(\mu) \psi_p \, dV , \tag{A4.15}$$

where $H(\mu)$ is the one-electron Hartree–Fock–Roothaan operator defined by the equation (A4.2). The total energy of all the electrons, E_{el}, corresponding to the determinantial wave function (A4.14) can be found from the relation

$$E_{el} = \int \Psi_A^* \mathscr{H}(\mu) \Psi_A \, dV. \tag{A4.16}$$

In this case the Hamiltonian $\mathscr{H}$ is equal to a sum of operators

$$\mathscr{H} = \sum_{\mu=1}^{N} H(\mu) \tag{A4.17}$$

and the total electronic energy of the molecule is

$$E_{el} = \tfrac{1}{2} \sum_i \sum_j P_{ij}(H_{ij}^{core} + H_{ij}) = \sum_i^{occ} E_i + \tfrac{1}{2} \sum_i \sum_j P_{ij} H_{ij}^{core}. \tag{A4.18}$$

The total energy of the system is the sum of the electron energy and the nuclear repulsion component

$$E = E_{el} + E_{nn}. \tag{A4.19}$$

The energy E_{nn} is most often calculated as a sum of Coulomb interactions.

The main difficulty in applying of the LCAO-MO method lies in the determination of the matrix elements given by (A4.9), since their calculation requires knowledge of the molecular wave function. For this reason, solution of the Hartree–Fock–Roothaan equations is carried out by an iterative method. Starting with approximate wave functions, matrix elements H_{ij} are found and then new functions which, as a rule, are better approximations. In the iteration process, increasingly correct values of the electron potential are determined at each step; thus this solution of equations (A4.7) is called a *self-consistent field method*, LCAO SCF [2].

Calculations based on the LCAO-MO-SCF method presented above with no additional approximations in the form of semiempirical parameters, are called *ab initio* calculations. The data obtained by this method are highly accurate, and in several cases the accuracy of such calculations is comparable to and may even surpass that of experimental results [3]. This is especially true for quantities relating to a molecule in its ground state, such as the total energy of the system, the dipole moment and (to a lesser extent) the ionization energy and conformation of molecules.

The physicochemical properties of molecules are much more correctly reproduced in the quantum theory exceeding the one-electron approximation of Hartree–Fock–Roothaan. Here belongs in the first place the *configuration interaction method* (CI), where the many-electron wave function is constructed in the form of linear combination of orthogonal configuration functions of the type (A4.14). This procedure [4] is especially useful in investigations of excited states and their characteristic quantities. Because of the complicated computational procedure for two-electron integrals, *ab initio* calculations are still restricted to relatively small molecules. For larger molecules, despite intense development of calculation procedures in recent years, it is necessary to use approximate methods. In these, it is possible to simplify considerably the form of the expression for the matrix elements (A4.9) by assuming a restricted number of integrals and introducing semiempirical parameters. Reduction of the number of integrals is possible, firstly using the Zero Differential Overlap (ZDO) approximation [5], according to which

$$\chi_i^* \chi_j = \delta_{ij} \tag{A4.20}$$

that is, integrals, for which $\chi_i^* \chi_j = 0$, when $j \neq i$, are neglected. This approximation significantly reduces the number of integrals needed, particularly the two-electron three- and four-centre integrals of the type (A4.12) and (A4.13) which are the most difficult to calculate.

A further substantial simplification of the *ab initio* method is obtained by reduction of the atomic basic set (A4.3) to the valence orbitals, or even to π-type orbitals. Such constraints, together with the relation (A4.17) leads to a series of relatively simple quantum computational methods.

1. VALENCE METHODS

a. NDDO approximation. In the LCAO-MO-SCF-NDDO (Neglect of Diatomic Differential Overlap) method, worked out by Pople *et al.* [6], two-electron integrals are omitted, when atomic orbitals with indices i, j or k, l belong to different centres. The expression for the matrix elements is then reduced to the relation

$$H_{ij} = H_{ij}^{\text{core}} + \sum_{B} \sum_{k,l \in B} P_{kl}(ij|kl) - \tfrac{1}{2} \sum_{k,l \in B} P_{kl}(ik|jl), \tag{A4.21}$$

when $i, j \in A$ (atomic orbitals with indexes i, j are on atom A), and

$$H_{ij} = H_{ij}^{\text{core}} - \tfrac{1}{2} \sum_{k \in A} \sum_{l \in B} P_{kl}(ik|jl), \tag{A4.22}$$

when $i \in A$ and $j \in B$, i.e. if atomic orbitals belong to different centres (A or B) of a molecule. The components H_{ij}^{core} are

$$H_{ij}^{\text{core}} = U_{ij} + \sum_{B \neq A} (i|V_B|j) \tag{A4.23}$$

and

$$U_{ij} = \left(i \left| -\frac{\hbar^2}{2m} \nabla^2 - \frac{Z_A e^2}{r_A} \right| j \right), \tag{A4.24}$$

while

$$V_B = -\frac{Z_A e^2}{r_B}. \tag{A4.25}$$

The NDDO method is the most advanced valence method but since there are still a significant number of two-electron integrals, not many calculations have been done by this method to date. It seems, however, that this method can be most useful in problems concerning the electron structure of molecules in their ground state.

b. INDO and MINDO approximations. In the LCAO-SCF-INDO (Intermediate Neglect of Differential Overlap) [7] and LCAO-MO-SCF-MINDO (Modified Intermediate Neglect of Differential Overlap) methods, the interelectronic repulsion integrals are taken into account only when atomic orbitals are on the same centre. According to this, the matrix elements H_{ij} can be written in the form

$$H_{ii} = U_{ii} + \sum_{k,l \in A} P_{kk}(ii|kk) - \tfrac{1}{2}(ik|ik) + \sum_{B \neq A} (P_{BB} - Z_B)(AA|BB), \tag{A4.26}$$

when $i \in A$, with $(AA|BB)$ denoting the two-centre Coulomb integral for ns atomic orbitals. Additionally,

$$H_{ij} = -\tfrac{1}{2}P_{ij}[(ii|jj) - 3(ij|ij)], \tag{A4.27}$$

when $i, j \in A$, and

$$H_{ij} = U_{ij} - \tfrac{1}{2}P_{ij}(AA|BB), \tag{A4.28}$$

if $i \in A$ and $j \in B$.

The INDO method gives correct values for the dipole moments, and also correctly predicts the geometry of simple molecules. It is used in investigations of electronic spectra [9]. In its open-shell version, correct values of spin densities, are also obtained, thus allowing an interpretation of the magnetic properties of molecules.

In contrast to the INDO approximation, in the MINDO method, integrals in the matrix elements H_{ij} are determined semi-empirically. This method allows extraordinarily good agreement to be obtained between computed and experimental results for quantities such as the ionization energies, heats of formation, rotational barrier heights and the geometry of molecules.

c. CNDO method. In the commonly-used LCAO-SCF-CNDO (Complete Neglect of Differential Overlap) [10,11] method, the two-electron integrals are considered only when $i, j \in A$ and $k, l \in B$. The Hartree–Fock–Roothaan matrix elements are then given by equations

$$H_{ii} = U_{ii} + P_{AA} - \tfrac{1}{2}P_{ii}(AA|AA) + \sum_{B \neq A}(P_{BB} - Z_B)(AA|BB), \qquad \text{(A4.29)}$$

$$H_{ij} = U_{ij} - \tfrac{1}{2}P_{ij}(AA|BB), \qquad \text{(A4.30)}$$

where integrals U_{ij} are determined by relation

$$U_{ii} = -\tfrac{1}{2}(I_i + E_i) - (Z_A - \tfrac{1}{2})(AA|AA). \qquad \text{(A4.31)}$$

I_i and E_i are the ionization energy and electron affinity of the i-th atomic orbital respectively and Z_A denotes the charge on the nucleus A. Non-diagonal matrix elements U_{ij} are approximated by the relation

$$U_{ij} = \tfrac{1}{2}(\beta_A^0 + \beta_B^0)S_{ij}, \qquad \text{(A4.32)}$$

in which β_A^0 and β_B^0 are semi-empirical parameters characterizing atoms A and B, and S_{ij} denotes the overlap integral. P_{AA} and P_{BB} in equation (A4.26) are total electron densities on atoms A and B, while P_{ii} is the electron density corresponding to the i-th atomic orbital.

The CNDO method, mostly known under the name CNDO/2, is used in many problems such as calculations of dipole moments, studies of intermolecular forces and conformational problems and gives a correct interpretation of a series of physicochemical properties of molecules. In its modified version [12] this method is also used in the studies of excited states.

d. Extended Hückel (EH) method. In the LCAO-MO-EH method [13–15] an effective operator H is introduced instead of the Hartree–Fock–Roothaan operator (A4.2)

$$H = H_{\text{eff}}, \qquad \text{(A4.33)}$$

so that the matrix elements H_{ij} are approximated by semi-empirical relations. Diagonal terms H_{ii} are usually assumed to be equal to the ionization energies for electrons of the corresponding valence orbitals of the atom

$$H_{ii} = -I_i, \qquad \text{(A4.34)}$$

while non-diagonal elements are approximated by the Wolfsberg–Helmholtz [16] relation

$$H_{ij} = \tfrac{1}{2}KS_{ij}(H_{ii}+H_{jj}) \tag{A4.35}$$

or the Ballhausen–Gray relation [17]

$$H_{ij} = KS_{ij}\sqrt{H_{ii}H_{jj}}\,. \tag{A4.36}$$

In both cases K is a semi-empirical parameter and S_{ij} is the overlap integral.

With the form of the operator assumed in equation (A4.33), the total electronic energy becomes equal to the sum of the energies E_p of occupied molecular orbitals

$$E_{\mathrm{el}} = \sum_{p}^{\mathrm{occ}} N_p E_p\,, \tag{A4.37}$$

where N_p denotes the occupation number of the p-th molecular orbital.

The **EH** method has been used in calculations of ionization energies, dipole moments, rotational barriers, and excited states even for large chemical molecules. Still better results are obtained using an iterative version (IEHM — Iterative Extended Hückel Method), in which the diagonal matrix elements H_{ii} are dependent on the charge q of the atom

$$H_{ii} = A_i + B_i q + C_i q^2. \tag{A4.38}$$

The constants A_i, B_i and C_i for a given type of atomic orbital are determined from the known electronic structure of the atom.

e. SCCC Method. The LCAO-MO-SCCC (Self-Consistent Charge and Configuration) approximation [18–20] is a simplified molecular orbital method, in which the diagonal elements of the Hartree–Fock–Roothaan matrix are dependent not only on the charge on the central ion, but also on its electron configuration. Non-diagonal elements H_{ij} are, as in the EH method, approximated by relations of the type (A4.35) or (A4.36). In order to determine the necessary integrals, semi-empirical parameters must be introduced and these are usually obtained from atomic spectroscopy data. Numerical calculations are then carried out until self-consistency is achieved for the assumed central ion charge and electron configuration.

The SCCC version of the molecular orbital method has been used in studies of the electronic structure of compounds containing transition metal atoms, where there is a substantial contribution from *d* or *f* electrons. In addition to crystal field and ligand field theories, this is an important method in studies of the structure of coordination compounds.

2. π-ELECTRON METHODS

These methods are based on a π-electron approximation [21], in which the Hamiltonian of the each of π-electrons in a molecule has the form of the Hartree–Fock–Roothaan operator (A4.2). The wave function of all the electrons in the system under consideration in then expressed as expressed as a product of the wave functions of σ- and π-electrons

$$\Psi = (\Sigma)(\Pi) \tag{A4.39}$$

and in this formalism the total electron energy of a molecule may be expressed as the sum

$$E_{el} = E_\sigma + E_\pi . \tag{A4.40}$$

In the π-electron model, based on the ZDO approximation (A4.17), only π-electrons are considered explicitly and their motion is investigated in the combined field due to the nuclei and the other electrons in the molecule [21].

a. Pariser–Parr–Pople method. In this method [21], the molecular orbitals are approximated by a linear combination (A4.3) of the atomic orbitals $2p_z$ of the molecule. Using the Hartree–Fock–Roothaan operator in the form (A4.2), it is possible to write the matrix elements H_{ij} as

$$H_{ii} = \alpha_i + \tfrac{1}{2}P_{ii}(ii|ii) + \sum (P_{jj} - Z_j)(ii|jj) , \tag{A4.41}$$

$$H_{ij} = \beta_{ij} - \tfrac{1}{2}P_{ii}(ii|jj) , \tag{A4.42}$$

where

$$\alpha_i = \int \chi_i^* H^{core} \chi_i \mathrm{d}V , \tag{A4.43}$$

$$\beta_{ij} = \int \chi_i^* H^{core} \chi_j \mathrm{d}V . \tag{A4.44}$$

The integrals (A4.43) and (A4.44) are usually determined by semi-empirical methods.

The Pariser–Parr–Pople method, combined with configurational superposition has been used in studies of the electronic spectra of compounds containing π-electrons. In most cases, both the results of calculations of electronic structure and of the intensity of the corresponding optical transitions represent an important supplement to experiment.

b. Hückel method. Neglecting the two electron integrals in the matrix elements (A4.41) and (A4.42) one obtains

$$H_{ij} = \alpha_i , \tag{A4.45}$$

$$H_{ij} = \beta_{ij} . \tag{A4.46}$$

This means that the LCAO-MO method in the Pariser–Parr–Pople approximation reduces to the Hückel method [22], known also as the HMO (Hückel Molecular Orbitals) approximation which has been applied for some time to organic molecules. In practice, however, instead of relations (A4.41) and (A4.42), semi-empirical relations of the form

$$\alpha_i = \alpha_C + k_i \beta_{CC} , \tag{A4.47}$$

$$\beta_{ij} = h_{ij} \beta_{CC} \tag{A4.48}$$

are used, where α_C is the Coulomb integral for a carbon atom and β_{CC} the resonance integral for the aromatic bond [23]. Neglecting integrals β_{ij} for non-neighbouring atoms, the properties of i-th atom in the molecule are determined by the parameter k_i, while the properties of the bond between a pair of atoms i–j are determined by the parameter h_{ij}. Numerical values of the k_i- and k_{ij}-parameters can be found in the relevant literature [23].

The Hückel method was used in the early days of the development of quantum chemistry, and it is now only a semi-quantitative model, giving explanations for many of the chemical properties of π-electron systems in a unified way.

3. MULTIPLE-SCATTERING $X\alpha$ METHOD

During the last decade there has been growing interest in the $X\alpha$ method conceived and formulated by Slater and Johnson [24–26]. Within this formalism, the entire space containing the molecule is geometrically divided into three regions and, according to the muffin-tin approximation, the electron density and the Coulomb potential of each region is spherically averaged with respect to the centre of the corresponding sphere. The first region is composed of spheres centred on the constituent atoms of the system with radii dependent on the atom type. The spheres touch each other but do not overlap. Inside the spheres the exact potential of an electron is replaced by its spherical average. The second region, surrounding the entire molecule, is called Watson sphere and the potential is assumed to be averaged over the interatomic region. Within the third spherical region beyond Watson sphere, the exact potential is also approximated by a spherical average.

For each of three regions mentioned above, a one-electron Schrödinger equation holds

$$\left| -\frac{\hbar^2}{2m} \nabla^2 + V_C(r) + V_{X\alpha}(r) \right| \psi(r) = E\psi(r) . \tag{A4.49}$$

Here, the first term is the standard kinetic energy operator for an electron and $V_C(r)$ denotes the Coulomb potential. Furthermore, the $X\alpha$ potential, is given according to the Slater statistical model of atom, by

$$V_{X\alpha}(r) = -6\alpha[\tfrac{3}{8}\pi\varrho(r)]^{1/3} , \tag{A4.50}$$

where α is an undetermined parameter (hence the name of the method — "X" refers to this indeterminancy) and $\varrho(r)$ defines the electron density at the point r. The solution of the equation (A4.49) gives the wave functions of the three spherical regions. The wave-function obtained in this approximation for the second region can be regarded as a sum of two terms, an incident and a scattered wave. The boundary conditions, which require continuity of the wave-functions and their derivatives across each spherical boundary, lead to the set of secular equations from which the eigenvalues E_p and the respective spinorbitals $\psi_p(r)$ can be evaluated.

In the iterative MS $X\alpha$ (Multiple Scattering) method the Schrödinger equation is solved for an assumed potential giving a certain number of spinorbitals. Choosing spinorbitals which are occupied in the order of increasing energies, the electron density may be determined from the relationship

$$\varrho(r) = \sum_p N_p \psi_p^*(r) \psi_p(r) , \tag{A4.51}$$

where N_p denotes the occupation number of the spinorbital $\psi_p(r)$. In the next step, the electron density can be used to determine the $V_{X\alpha}(r)$ potential (A4.50). This is then used as the starting point for the next iteration. Usually the starting potential for this kind of SCF procedure is generated by a superposition of atomic electron densities where the atomic basis set may be built from analytical Hartree–Fock orbitals in either STO (Slater Type Orbitals) or GTO (Gaussian Type Orbitals) version.

The $X\alpha$ method has been used with some success for the evaluation of ground state properties as well as for calculation of optical transition energies in relatively complicated molecular systems — including coordination compounds and metallic aggregates. Unlike most other SCF-type methods, long-range interelectron correlation is automatically included so that the results appear to be in better agreement with experiment than those of the standard LCAO-MO-SCF scheme. This approach is also approximately two orders of magnitude faster than the molecular orbital method. However, a unified scheme for the choice of spherical radii does not yet exist. It should be noted that in ordered metallic systems, the wave function of the third region should satisfy the Bloch condition in going from one unit cell to another.

REFERENCES

[1] C. C. J. Roothaan, *Rev. Mod. Phys.*, **23**, 69 (1951).

[2] I. G. Csizmadia, M. C. Harison, J. W. Moskowitz and B. T. Sutcliffe, *Theoret. Chim. Acta*, **6**, 191 (1966).

[3] W. Kołos and C. C. J. Roothaan, *Rev. Mod. Phys.*, **32**, 219 (1960).

[4] B. Roos, *Computational Techniques in Quantum Chemistry and Molecular Physics*, Eds. G. H. F. Diercksen, B. T. Sutcliffe and A. Veillard, Reidel Publishing Co., Dordrecht–Boston 1975.

[5] I. Fischer-Hjalmars, *J. Chem. Phys.*, **42**, 1962 (1965).

[6] J. A. Pople, D. P. Santry and G. A. Segal, *J. Chem. Phys.*, **43**, S129 (1965).

[7] J. A. Pople, D. L. Beveridge and P. A. Dobosh, *J. Chem. Phys.*, **47**, 2026 (1967).

[8] N. C. Baird and M. J. S. Dewar, *J. Chem. Phys.*, **50**, 1262 (1969).

[9] J. Lipiński, A. Nowak and H. Chojnacki, *Acta Phys. Polon.*, **53a**, 229 (1978).

[10] J. A. Pople and G. A. Segal, *J. Chem. Phys.*, **44**, 3289 (1966).

[11] J. A. Pople and D. L. Beveridge, *Approximate Molecular Orbital Theory*, McGraw-Hill, New Nork 1970.

[12] J. Del Bene and H. H. Jaffé, *J. Chem. Phys.*, **48**, 1807 (1968).

[13] C. J. Ballhausen and H. B. Gray, *Molecular Orbital Theory*, W. A. Benjamin, Inc., New Nork–Amsterdam 1965.

[14] D. A. Brown, W. J. Chambers and N. J. Fitzpatrick, *Inorg. Chim. Acta*, **6**, 7 (1972).

[15] R. Hoffmann, *J. Chem. Phys.*, **39**, 1397 (1963).

[16] M. Wolfsberg and L. Helmholtz, *J. Chem. Phys.*, **20**, 837 (1952).

[17] C. J. Ballhausen and H. B. Gray, *Inorg. Chem.*, **1**, 111 (1962).

[18] A. Viste and H. B. Gray, *Inorg. Chem.*, **3**, 1113 (1964).

[19] H. Bash, A. Viste and H. B. Gray, *J. Chem. Phys.*, **44**, 10 (1966).

[20] A. Gołębiewski, *Quantum Chemistry of Inorganic Compounds*, PWN, Warsaw 1969 (in Polish).

[21] R. G. Parr, *The Quantum Theory of Molecular Electronic Structure*, W. A. Benjamin, Inc., New York–Amsterdam 1964.

[22] F. A. Cotton, *Chemical Applications of Group Theory*, 2nd edition, Wiley-Interscience a Divison of John Wiley and Sons, Inc., New York–London–Sydney–Toronto 1971.

[23] A. Streitwieser, Jr., *Molecular Orbital Theory*, J. Wiley, New York–London 1961.

[24] J. C. Slater, *Adv. Quantum Chem.*, **6**, 1 (1972).

[25] J. C. Slater and K. H. Johnson, *Phys. Rev.*, **5B**, 844 (1972).

[26] K. H. Johnson, *Adv. Quantum Chem.*, **7**, 143, 1973).

5

Interaction of Small Metallic Particles with a Non-metallic Substrate and its Importance for Composite Catalysts and Thin Films

Small metallic particles are almost always found as a dispersion in a non-metallic medium, or are deposited on the surface of the non-metallic substrate. Due to their interaction with the substrate they are not able to move freely. The forces of this interaction are in most cases strong enough although they can be of different nature, depending on the kind of metal and support, and their magnitude can vary in broad limits. In what follows, the results of investigations into these interactions are presented.

As one might expect, there are no experimental methods which provide a direct determination of the strength of the interaction between single atoms, or between small aggregates of atoms and a substrate. Thus it is necessary to estimate the magnitude of such interactions from experimental data for particles of sufficiently large dimensions, or from macroscopic interfaces between a metal and its substrate.

5.1 DETERMINATION OF THE MAGNITUDE OF THE INTERACTION BETWEEN METAL AND SUBSTRATE BASED ON MEASUREMENTS OF THE CONTACT ANGLE

One of the quantities which characterize the interface between two phases and which is accessible to direct measurement, is the contact angle — that is, the angle formed by a drop of a liquid phase lying on a solid substrate. Generally, when a metallic droplet solidifies, its shape does not change and thus the contact angle retains its meaning for the solid phase also.

The relation between the contact angle, γ, and the free surface energies, σ_{ij}, at the metal–substrate, substrate–vacuum (gas) and metal–vacuum phase boundaries is expressed by the Young equation (see Fig. 5.1),

$$\cos\gamma = \frac{\sigma_{2,3}-\sigma_{1,3}}{\sigma_{1,2}}, \tag{5.1}$$

where $\sigma_{i,j}$ is the surface tension at the boundary between phases i and j.

The work of adhesion, P_{adh}, that is the energy needed to separate the metal from the substrate is, according to Dupré,

$$P_{adh} = \sigma_{2,3}+\sigma_{1,2}-\sigma_{1,3}\,. \tag{5.2}$$

After substituting (5.1) this becomes

$$P_{adh} = \sigma_{1,2}(1+\cos\gamma)\,. \tag{5.3}$$

A sharp contact angle (positive $\cos\gamma$, or $\gamma<\pi/2$) occurs when the difference $\sigma_{2,3}-\sigma_{1,3}$ is positive. There is then a tendency to diminish the free surface energy of the solid 3 by the liquid phase *1* leading to wetting.

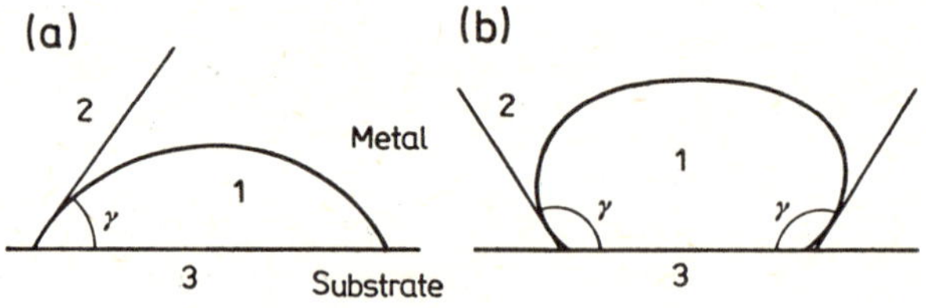

Fig. 5.1 — Contact angles: (a) sharp, (b) obtuse.

With an obtuse contact angle, $\cos\gamma$ is negative and the difference $\sigma_{2,3}-\sigma_{1,3}$ is negative. In the limiting case when $\gamma = \pi$, phase *1* forms a spherical drop on the surface of the phase *3* (no wetting) — see Fig. 5.1.

The free surface energy (surface tension) is not easy to determine experimentally and few attempts have been made to date. Humenik and Kingery [1], and Kingery [2] determined the contact angle and the interfacial energy of liquid iron and nickel on a series of ceramic oxides (Al_2O_3, MgO, ZrO_2) using the sessile drop method. Piliar and Nutting [3] have performed similar measurements for solid metals (Au, Ag, Cu, Ni and γ-Fe on Al_2O_3). They found that the interfacial energy of fcc metals with $\{111\}$ planes orientated towards the substrate differs only slightly from that of the liquid metal on the same substrates. Thus it seems likely that no substantial change in the energy of adhesion occurs at the freezing temperature of these metals. Adhesion energy values of 0.2 to 0.9 $kJ \cdot m^{-2}$ were observed in the cases mentioned and taking into account the surface density of atoms (for Ni about 1.8×10^{19} $atoms \cdot m^{-2}$) the energy of adhesion to the ceramic substrate is found to be 7–30 $kJ \cdot g\ atom^{-1}$.

H. de Bruin *et al.* [4] in their studies of metal–ceramic bonds have measured the work of adhesion of layers of metals on MgO, Al_2O_3 and glass by measuring the shearing force necessary to break the bond. In most cases they found that the shear strength of the interface exceeds that of the ceramic or glass substrate.

5.2 EXPERIMENTAL DETERMINATION OF THE ADHESION OF THIN METAL LAYERS

Benjamin and Weaver [5, 6] have developed a direct quantitative method for the measurement of the strength of the adhesion between metallic layers and a substrate. This method is based on an established procedure used to compare adhesion in different metal-substrate systems and consists of drawing a rounded stylus across the layer. The load on the stylus is increased and at a critical load, the stylus reaches the bottom of the layer and tears it from the substrate. This method gives a reasonable estimate of the adhesion of a given metal to the substrate. The measured shearing force exerted on the layer, is

$$F_{sh} = \frac{aP}{\sqrt{r^2 - a^2}}, \qquad (5.4)$$

where P is the hardness of the substrate, a is the width of the furrow in the layer, equal to $a = \sqrt{W/\pi P}$, W is the weight on the stylus and r is the radius of curvature of the stylus tip (typically 10^{-2}–10^{-3} cm).

For small loads and with $a \ll r$

$$F_{sh} = \sqrt{\frac{WP}{\pi r^2}} \,. \tag{5.5}$$

The critical load W is, over a wide range, independent of the mechanical properties of the layer and of the substrate and the layer thickness. This indicates that the shearing force is related only to the interaction between atoms in contact with atoms of the substrate. In this way, Benjamin and Weaver were able to show that adhesion of metals such as Au, Ag, Al, Cu, Zn, Cd on alkali halide substrates, NaCl, KCl and KBr, can be completely explained by van der Waals forces. Layers of the above metals on alkali halides do not show any ageing, that is, the adhesion does not change with time and is independent of the atmosphere (nature of the gas and its pressure) in which they are formed and stored.

The magnitude of the van der Waals forces can be calculated theoretically using the formula given by Benjamin and Weaver [7]. Restricting these interactions to London forces only

$$E_{adh} = \frac{N\alpha\alpha' I I'}{4r^3(I+I')} \,, \tag{5.6}$$

where N is the number of atoms of the metal corresponding to the unit volume of substrate, α and α' are polarizabilities of atoms of the metal and of the substrate respectively, r is their mutual distance, and I, I' are ionization potentials of the metal and substrate atoms respectively.

Values of the adhesion energy of thin films of Al, Ag and Cd on glass, calculated in this way, are 8.4, 11.5 and 14.5 $J \cdot m^{-2}$, respectively and 11.5 and 10.5 $J \cdot m^{-2}$ for Ag on the (100) planes of NaCl and KBr. These values are in good agreement with the experimentally measured values of the condensation energy which is obtained by measuring the pressure of the metal vapour at the critical condensation temperature.

Methods for estimating the adhesion based on an examination of the resistance to abrasion, or on measurements of the perpendicular force needed to strip the film from the substrate, which had been used earlier, give only qualitative and sometimes irreproducible results. In the case of the abrasion technique, the result is also dependent on the hardness of the metal layer, and a determination of the force needed to separate the metallic layer from the substrate is possible only when adhesion between the layer and the test rig is higher than that of the layer to the substrate. For these reasons, these methods are now only of historical importance.

5.3 THE INFLUENCE OF REACTIVE GASES ON THE ADHESION

During the course of investigations on the adhesion of metals to glass and ceramic substrates, it has been noted that the adhesion of metallic layers is increased, when the films are formed in an atmosphere of oxygen, or stored in an atmosphere containing oxygen. These observations have been confirmed and quantified to some extent by Benjamin and Weaver. Applying methods described for measuring the strength of adhesion, these authors showed that noble metals, such as silver and gold exhibit characteristically weak adhesion to glass in a vacuum of the order of 10^{-3} Pa. Moreover, exposure to oxygen after evaporation does not bring about any increase in adhesion. Films of magnesium and zirconium on the other hand, which have a great affinity for oxygen form a strong bond even after evaporation in vacuum, while molybdenum, titanium, iron, chromium and nickel display weak adhesion immediately after evaporation, which increases rapidly on exposure to oxygen. Apart from the affinity for oxygen, the structure of the film and its porosity plays a substantial role, modifying the extent to which oxygen can diffuse to the metal–glass interface. Thus low-melting metals like zinc, cadmium, tin and lead, which form compact films, exhibit weak adhesion which does not increase in an oxygen atmosphere. Similarly, there is no oxygen-induced increase in adhesion for aluminium films, which are covered with a thin, compact Al_2O_3 layer which is impermeable to oxygen.

Based on these facts, it has been assumed that the most crucial factor in the increase of adhesion of metals to glasses and ceramics is their ability to form layer of oxide on the metal–substrate interface [9]. When, for instance, a layer of a non-stoichiometric oxide with the composition close to Bi_2O_3 is formed on glass, the adhesion of the gold film subsequently deposited is greatly increased. Also, if gold is deposited on glass by cathodic sputtering at a partial pressure of oxygen of about 1 Pa, the resulting gold film exhibits strong adhesion, which decreases when the film is heated in vacuum and again increases after heating in oxygen [9]. The influence of oxygen on the strength of the metal–glass interfacial bond has been demonstrated by the experiments of Pask and Fulrath [10], carried out with glass drops in contact with a metallic substrate. They showed that there are changes in the contact angle in the presence of oxygen — even for gold which is considered to be the most noble metal. The increased adhesion between gold and quartz in the presence of oxygen has also been reported by other authors [11].

5.4 STRONG INTERACTIONS OF METALS WITH THE SUBSTRATE

The above examples of strong adhesion of metal to the substrate in an oxygen-containing environment suggest the possibility of an intermediate layer of oxide. More recent investigations however, show that there are many cases where such strong interactions exist in the absence of intermediate layers, both in vacuum and in the presence of reactive gases. Most of the observations pertinent to the strong interaction between metals and non-metallic substrates have been made on samples in which the metal is very highly dispersed.

Strong interactions between a metal and a non-metallic substrate may be due to "chemical" forces, connected with the transfer of electrons from metal to the support or vice versa. The work of Hamilton and Logel [12–14] on the interaction of silver and palladium with a carbon substrate and that of Lewis [15] dealing with silver/carbon and the bismuth/carbon systems clearly show the high interaction energy of these metals with the substrate: for silver on carbon it is equal to 92 $kJ \cdot g\ atom^{-1}$. Such strong interactions can also be confirmed by theoretical calculations, such as those based on quantum chemical methods. The work of Baetzold, mentioned in Chapter 4, on clusters of Pd and Ag on C represent an example [16, 17].

There are cases, when the forces of interaction with the substrate are comparable, and even greater than those between the metal atoms in the film. In such cases, the early stages of condensation of metal vapours on the substrate is marked by the appearance of flat, quasi two-dimensional structures rather than three-dimensional aggregates. Such two-dimensional films of metal ("raft" and "pillbox structures") also tend to form when the metal is deposited by chemical reduction, as for example, rhodium chemically deposited on Al_2O_3 [18] and SiO_2 [19].

5.4.1 Strong interactions in an atmosphere of a reactive gas

Metal–substrate–gas systems are far more common in practice than the metal-substrate system typical of a "good" vacuum (i.e. at a pressures below, say, 10^{-7} Pa). Reactive gases such as oxygen, hydrogen, water vapour, carbon monoxide, etc. usually influence the strength of interaction between metallic particles and the substrate to such an extent that the size, shape and mobility of the particles at elevated temperatures are altered.

The basic method for observing changes in these properties is electron microscopy. Changes in the shape of particles are easily observed, for instance the formation of "flat" structures due to a decrease in the contact angle resulting from an increase in the strength of the interaction, as well as a decrease in mobility. Observations have been mostly performed with systems constituting, or related to, supported metal catalysts, whose properties such as degree of dispersion, chemical composition and particle shape are of great importance in catalytic activity. Moreover, changes in these properties at higher temperatures and in the presence of reactive gases directly influence their stability during the reaction.

Systems which model supported platinum catalysts have been investigated over a wide temperature range in both reducing (hydrogen) and oxidizing atmospheres. At relatively high temperatures (773–1123 K) a decrease in the interaction of small platinum particles (under 10 nm) was observed on Al_2O_3 substrates after alternate exposure to hydrogen and oxygen. This was shown by an increase in diffusional mobility, whilst in an atmosphere of hydrogen containing water vapour, an increase in crystallite size occurred. In pure hydrogen however, a strong interaction between platinum and the Al_2O_3 support leads to the formation of the intermetallic compound Pt_8Al_{21} [20].

Irregularities and defects increase the interaction between the metal and an oxide support, as observed in the case of small palladium particles [21].

Strong chemical interactions, are more likely to occur in an atmosphere of hydrogen and on substrates which are more easily reduced by hydrogen than Al_2O_3. This has been demonstrated in investigations of the platinum/TiO_2 system [22,23], where it has been established that platinum at 875 K in an atmosphere of hydrogen forms quasi-flat, highly dispersed particles, and that the TiO_2 is reduced to a lower oxide, Ti_2O_7. An increase in the metal–support interaction seems to be connected with changes in the chemical composition of the substrate. Similar phenomena have also been observed for the Ni/TiO_2 system [24], and for evaporated iron films on TiO_2, for which the strength of the chemical interaction is shown by the formation of an iron–titanium alloy at a temperature of 875 K in hydrogen. In an oxidizing atmosphere the compound $FeTi_2O_5$ was formed [25].

In general, strong interactions between metals and oxide supports at high temperatures and in a reactive gas atmosphere can be attributed to chemical changes in the metal, the oxide, or both, occurring at the metal/oxide interface.

5.5 ELECTRICAL PHENOMENA AT THE METAL–SUBSTRATE INTERFACE

Strong interaction between a metal and a substrate with no chemical interaction (vacuum environment, low temperature) have been interpreted as a shift of electrons from the metal to the substrate, or vice versa. The direction of such a shift is determined by the difference between the Fermi levels (work functions) in each phase. A transfer of electrons is accompanied by the appearance of a space charge at the phase boundary. To describe this phenomenon and its consequences, theories of the metal–semiconductor interface [26, 27] can be applied since a majority of substrates on which dispersed metals are deposited, have electrical properties characteristic of *n*-, or *p*-type semiconductors.

Since the transfer of charge at the phase boundary is restricted to distances of, at most, a few tens of nanometres, direct measurement of the shift is difficult, if not impossible. In the case of thin metal films on the surface of a semiconductor, or vice versa, changes connected with the transfer of charge can be observed. For example, modification of the electronic structure of one of constituents may be followed by observing changes in catalytic properties, which are believed to depend on the electron concentration in the catalytically active phase.

Two of the cases mentioned above: metals on a *p*- or *n*-type semiconductors have been experimentally studied by Schwab and co-workers. In the first, nickel was supported on an Al_2O_3 substrate, the *n*-type properties of which were modified by addition of divalent oxides thus lowering the electron concentration in the Al_2O_3, or tetravalent metal oxides which increased the electron concentration. An increase in the density of electrons in the Al_2O_3 carrier produced an increase in electrical conductivity and a simultaneous decrease of the catalytic activity of the nickel.

The magnetization of nickel measured in these samples showed a decrease, indicating a reduction of the magnetic moment of nickel, due to an increase in the *d*-band electron concentration [28]. The effect of an increase in the electron density of the support (Al_2O_3) is thus to transfer electrons to the metal, as indicated by the change of magnetic and catalytic properties of the latter.

Electron shifts in the reverse direction have been observed by the same author for the case of thin semiconducting films of NiO, Fe_2O_3 and ZnO supported on metallic silver. The work function of these semiconductors is higher than that of silver and electrons can flow from metal to the semiconductor, resulting in a decrease of the catalytic activity of the *p*-type semiconductor (NiO) and an increase for the *n*-type semiconductors (Fe_2O_3 and ZnO). In the case of zinc oxide, this effect was observable only for layer thickness below 50 nm, clearly indicating the restricted range of the charge transfer [29].

Investigation of a similar type have been carried out by Baddour and Deilbert [30] for thin metal films deposited on germanium, which is an intrinsic semiconductor. They established that the change in the catalytic activity of nickel in the decomposition of formic acid is connected with electron transfer.

Electron transfer from platinum to a carbon support has been postulated by Nicolau and Thom [31]. The evidence was obtained from electron spin resonance measurements of Pt/C catalysts, submitted to appropriate thermal treatment.

It has been often suggested that adhesion between a metal and its support is principally electrostatic in nature [32, 33]. Charge transfer produces an electrical double layer at the phase boundary, and the resulting electrostatic forces may be appreciable. A complete electrostatic theory of adhesion has been recently reported by Derjaguin and Smilga [34].

5.6 THE INTERACTION BETWEEN A METAL AND A SUBSTRATE AND ITS EFFECT ON THE GROWTH OF PARTICLES

Systems composed of very small metal particles deposited on a substrate — for example contact catalysts or thin metallic films — are thermodynamically unstable. At high, or even at moderate temperatures, a particle attempts to achieve thermodynamic equilibrium by minimizing the free energy of interfaces by increasing its size.

Growth of particles occurs by the transport of metal on the surface of the substrate. Based on theoretical considerations and experimental data, two mechanisms for transport and growth can be distinguished:

(a) surface diffusion by random movement of an entire particles followed by their subsequent collisions and coalescence — known as Smoluchowski's model,

(b) detachment of small fragments — normally atoms from the smaller particles, followed by surface migration and subsequent capture by larger particles — known as Oswald ripening.

5.6.1 Migration and coalescence of particles

When the interaction between the metal and the substrate is relatively weak, translational and rotational motion of metal particles a few nanometres in size becomes possible at temperatures of a few hundred degrees Kelvin. Such motion has been observed, for instance, in island films of gold on silica and carbon [35] and in aluminium, tantalum and cobalt films on a carbonaceous substrate at temperatures between 900–1050 K [36]. In other cases, indirect evidence of such movements, has been obtained by comparison of the observed rate of particle growth with that obtained under the assumption of a diffusion and coalescence growth process. However, these observations were made under conditions where the particles could chemisorb oxygen and catalytically oxidize carbon to CO and CO_2. Consequently, it seems likely that these secondary reactions could substantially influence the observed mobility of metal particles.

The first theoretical model for the formation of large particles from an ensemble of smaller ones, all with equal volumes, was described by Smoluchowski [37] for the specific case of the coagulation of colloidal solutions. In this model, small particles, surrounded by a "sphere of attraction" of radius R, collide due to Brownian motion. The mean square displacement, $\bar{x}^2$ is given by Einstein's relationship

$$\bar{x}^2 = 2Dt\,, \tag{5.7}$$

where D is the diffusion coefficient. Introducing the "diffusion time", T, equal to

$$T = \frac{1}{4\pi R D \nu_0}\,, \tag{5.8}$$

where ν_0 is the initial number of small particles in unit volume, Smoluchowski obtained an expression for the maximum number of particles of k-fold volume ν_k, appearing after a characteristic time $t = \frac{1}{2}(k-1)T$, i.e. after $k-1$ collisions,

$$\nu_k = 4\nu_0 \frac{(k-1)^k - 1}{(k+1)^k + 1}\,. \tag{5.9}$$

The quantity $\frac{1}{2}(k-1)T$ increases, of course, with increasing k, i.e. with the growth time of the particles.

Smoluchowski's coagulation theory has been further developed more recently by Fair and Gemmel [38]. A modification of this theory and extension to two-dimensional cases such as particles on a solid substrate was reported by Ruckenstein and Pulvermacher [39]. They described the growth of metallic particles in supported catalysts as a result of two processes; migration of the metal crystallites on the surface of the support, followed by collision and coalescence.

Diffusion of particles as a whole can occur not only by translation of the entire particle but also by the diffusion of surface atoms in a preferred direction. Atoms therefore accumulate on one side of the circumference of a particle leading to preferential growth on that side (see Fig. 5.2). This is clearly equivalent to movement of the particle in the

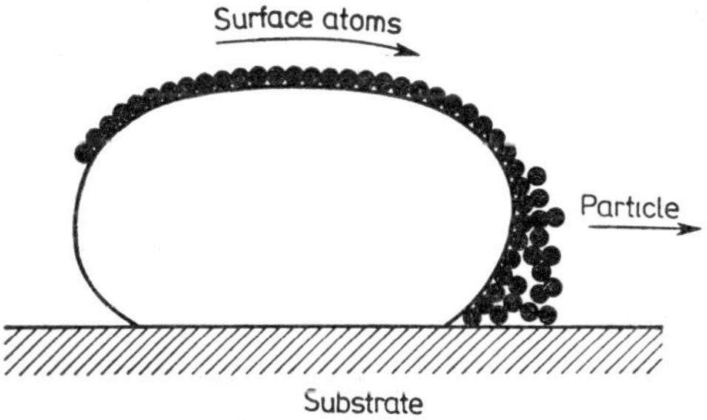

Fig. 5.2 — Diffusion of a particle on the substrate by surface atom transport.

direction of preferred surface diffusion, and it has been shown that this process is also possible in polyhedral particles [40]. The mechanism requires significant anisotropy of surface diffusion, of course, and such anisotropy has indeed been found experimentally in certain cases [41]. For example, diffusion coefficient for the (110) plane of nickel at 1073 K differs by an order of magnitude in two perpendicular directions. There are other experimentally investigated cases which indicate significantly different values of surface atomic diffusion coefficients on different planes of both fcc [42] and bcc [43] metals, thus supporting the possibility of anisotropic surface diffusion.

The diffusion coefficient of a spherical particle can be calculated from a relation obtained by Gruber [44]

$$D = 4.816 D_{at}(a/d)^4, \tag{5.10}$$

where D_{at} is the atomic diffusion coefficient, a is the diameter of the atom and d is the diameter of the particle.

After collision of two particles, coalescence can occur very quickly, i.e. over times short compared to the time of migration on the substrate. In such cases, the overall rate of growth will be determined by the rate of the migration process — that is the growth is diffusion-limited. Conversely, in cases where coalescence of particles after contact is slow, in comparison with the speed of migration, the rate of the process will be coalescence-limited.

Coalescence by atomic surface diffusion has been described quantitatively by the model of Nichols and Mullins [45]. These authors considered the coalescence of two equal spherical particles and calculated the relaxation times for three distinct stages of this process:

(1) formation of a "neck" between particles,

(2) the disappearance of the "neck" with the formation of a single elongated particle,

(3) transformation of the new particle to a spherically shaped particle (see Fig. 5.3).

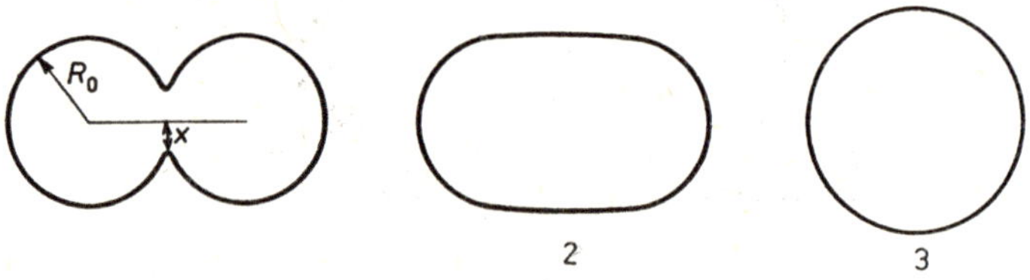

Fig. 5.3 — Three stages of the coalescence of identic spherical particles.

The corresponding expressions for the relaxation times are

$$\tau_{1(x/R=0.3)} = 2.92\times 10^{-5}\frac{R_0^4}{B} \quad \text{(stage 1)}, \tag{5.11}$$

$$\tau_2 = 0.166\frac{R_0^4}{B} \quad \text{(stage 1+2)}, \tag{5.12}$$

$$\tau_3 = 0.890\frac{R_0^4}{B} \quad \text{(stage 1+2+3)}. \tag{5.13}$$

Here, R_0 is the radius of the spherical particle before coalescence and

$$B = D_s N_0 \sigma_0 \Omega^2/kT, \tag{5.14}$$

where D_s is the diffusion coefficient for atomic migration on the surface of the particle, N_0 is the average density of adsorption sites on the particle surface, σ_0 is the average surface energy of the particle and Ω is the atomic volume of the particle.

Ruckenstein and Pulvermacher [39] proposed a physical model and kinetic equations for the growth of metal particles on a non-metallic support, assuming the growth to be brought about by diffusion of the metal particles consisting of multiples of a primary unit. Thus, particles with $k-i$ and i primary units merge and sinter after contact, to give particles consisting of k primary units. Taking into account the decrease in the number of particles composed of k units by the collisions with other particles, an expression is obtained for the change in the number of "k-fold" particles, ν_k with time (on a homogeneous surface)

$$\frac{d\nu_k}{dt} = \frac{1}{2}\sum_{i+j=k} k_{ij}\nu_i\nu_j - \nu_k \sum_{i=1}^{\infty} k_{ik}\nu_k\,, \tag{5.15}$$

where k_{ij} are the second order rate constants, dependent upon the mobility of particles on the substrate and on the nature of the interaction between the particles. For the case of diffusion-limited growth, these constants are equal to

$$k_{ij} = \frac{4\pi D_{ij}}{\ln 4T}\,, \tag{5.16}$$

where D_{ij} is the diffusion coefficient of the particle i with respect to particle j: $D_{ij} = D_i + D_j$. In an analogous way, the quantity $R_{ij} = R_i + R_j$, the radius of a merging pair of particles, was introduced. For the coalescence-limited growth, the following relation holds:

$$k_{ij} = 2\pi R_{ij}\alpha_{ij}\,, \tag{5.17}$$

where α_{ij} is a rate constant for the merging process. The quantity T in (5.16) is a dimensionless "time", corresponding to a diffusion time

$$T = \frac{D_{ij}\theta'}{R_{ij}^2}\,, \tag{5.18}$$

where θ' is the "small scale time" introduced by the authors to calculate the rate of collision of the particles, and the rate constants k_{ij}.

The actual surface of a support, on which the particles diffuse, is rarely homogeneous, both topographically and energetically. Thus there are surface sites, where the interaction between the support and the particles is strong enough to halt and capture the particles. The number of captured particles, ν_F, changes with time according to

$$\frac{d\nu_F}{dt} = (1-q)\sum_{i+j=k} k^*_{ij}\nu_i\nu_j + \tfrac{1}{2}(1-p)\sum_{i+j=k} k_{ij}\nu_i\nu_j - \nu_k\sum_{i=1} k_{ik}\nu_i + \beta_k\nu_k . \quad (5.19)$$

Here, p is a probability for the formation of an immobile particle by the collision of two mobile ones, q is the probability of formation of a mobile particle by the collision of immobile and mobile particles (for the sake of simplicity, it is assumed that both probabilities are independent of the size of the particles), β_k is the rate of immobilization of particles at sites with the strongest interaction.

If immobilization of particle is taken into consideration, the rate equation (5.15) becomes

$$\frac{d\nu_k}{dt} = \tfrac{1}{2}p\sum_{i+j=k} k_{ij}\nu_i\nu_j - \nu_k\sum_{i=1}^{\infty} k_{ik}\nu_i - \nu_k\sum_{i=1}^{\infty} k^*_{ki}\nu_i + q\sum_{i+j=k} k_{ij}\nu_{Fi} - \beta_k\nu_k \quad (5.20)$$

with constants k^*_{ij} appropriately modified with respect to k_{ij}.

Equations (5.15) and (5.16) were applied to the determination of the distribution of particle sizes as a function of time on a homogeneous surface. Based on these equations it is also possible to calculate the time-dependence of the accessible surface of the metallic particles, S, per unit surface area and of the substrate

$$\frac{dS}{dt} = -C_m S^m . \quad (5.21)$$

Here, constants C_m are functions of the parameters k_{ij} and D_{ij} (which in turn depend on the radii of the colliding particles). The exponent m for the case of diffusion-limited growth is equal to 4 or more, while for coalescence limitation, m lies between 2 and 3. Similarly, the value of C_m depends on k_{ij} and α_{ij}, which, in general, are functions of the radii of the colliding particles. It has subsequently been shown that relations of the type (5.21) hold for an arbitrary initial particles size distribution.

Relations between the size distribution and time can be obtained also for the case of an inhomogeneous substrate surface. Due to the presence of deep trapping sites, a fraction of the particles will be immobilized during the formation of the metal–substrate system. This leads to the possibility of reaching an equilibrium value of S during the growth process, whereas for the case of a homogeneous substrate, S decreases continuously. The equilibrium size distribution is strongly dependent on the initial distribution in contrast to the distribution on a homogeneous surface.

Ruckenstein and Pulvermacher have established that equation (5.21) adequately describes much of the experimental data relating to growth of platinum particles in catalysts supported on alumina in the temperature range 773–1073 K. The value of m occurring in the equation is about 2 indicating coalescence limitation, but for other temperature regimes m lies in the range 6–8, suggesting diffusion limitation.

Wynblatt and Gjostejn [46] have investigated the growth of platinum particles prepared by vacuum evaporation on an Al_2O_3 support at a temperature of 973 K in the presence of air or nitrogen with 2% addition of oxygen and found that the system does not obey kinetics based on a surface diffusion mechanism. These authors have concluded that, in these conditions, another growth law prevails, characterized by an increase in the exponent m in eq. (5.21) with time. However, the growth of platinum crystallites, supported on graphitized carbon in an atmosphere of nitrogen or hydrogen, can be adequately described with the aid of the Smoluchowski surface diffusion mechanism [47].

5.6.2 Ostwald ripening

"Ostwald ripening" is a mechanism for the growth of solid particles in which atoms (or molecules, etc.) become detached from the smaller particles and are captured by the larger ones. This accords with the known thermodynamic principle that smaller particles have a higher solubility (or vapour pressure) than larger ones. This is expressed quantitatively by the Gibbs–Thomson equation

$$\ln \frac{s}{s_a} = \frac{2M\sigma}{RT\varrho a} . \tag{5.22}$$

Here M, R, T are the molecular weight, gas constant and absolute temperature, respectively, s is a "normal" solubility, s_a is the solubility of a particle of radius a, ϱ is the density of the solid, and σ is its surface tension. This growth mechanism has general validity and was applied by Ostwald, for the first time to the recrystallization of precipitates in solutions [48].

The change of the free surface area of particles with time can be described, both for the case of Ostwald ripening and for the Smoluchowski mechanism, by the formula

$$\frac{1}{S^n} = \frac{1}{S_0^n} + kt , \tag{5.23}$$

where S and S_0 are the metal surface areas per unit surface area of the substrate, at time t, and zero time t_0, respectively. For the crystallite diffusion mechanism, the constant k is directly proportional to the volume of the metal per unit surface area of the substrate, and for Ostwald ripening it is independent of this volume. Exponents n in this equation (5.23) have values from 2 to 5 and are lower (by about 1) than for the corresponding processes in equation (5.21). The number of particles, in the system, N_t, present after time t is

$$N_t = N_{t_0} \frac{1}{1+(t-t_0)/t_0} . \tag{5.24}$$

The analysis of Ostwald ripening for the case of discontinuous metallic films, with a rate limited by diffusion of atoms on the substrate, has been carried out by Chakraverty [49]. The resulting relation between the so-called critical radius of the particle and the time of growth is

$$r_k = r_{k_0} \left[1 + \frac{t}{t_s}\right]^{1/4} . \tag{5.25}$$

Here r_k and r_{k_0} are critical radii, that is, radii at which a particle neither grows, nor decays, for times $t = 0$ and t. Critical radii can be assumed to be equal to the mean radii at the begining of the process and after a time t. The constant t_s is independent of the quantity of metal per unit surface area of the support, but inversely proportional to r_{k_0}.

Ostwald ripening has been used to describe the growth of solid uranium particles in liquid metallic sodium and lead [50]. The rate of change of size of the uranium particles due to diffusion of atoms from smaller to larger particles, was found to be

$$\frac{da}{dt} = DS \frac{2M\sigma}{RT\varrho^2 a}\left(\frac{1}{a_n} - \frac{1}{a}\right), \tag{5.26}$$

where D is the diffusion coefficient of the dissolved uranium, S is the solubility and a_n is the arithmetic mean of the particle diameters

$$a_n = \frac{1}{n}\sum_{i=1}^{n} a_i .$$

A number of published papers report applications of the Ostwald ripening model to the growth of metallic particles, on a non-metallic substrate. Some describe experiments in which active gases were present during high temperature annealing, which could react with the metal and in this way influence the transport of metal between particles. An example is the growth of platinum particles in the size range 150–300 nm at 1173 K which obeyed the Ostwald mechanism [51]. However, because of the presence of oxygen, there is some doubt whether the diffusing entities were platinum atoms, oxide molecules, or even ions.

A critical examination of the mechanism of metal particle growth on oxide substrates carried out by Flynn and Wanke [52] noted the discrepancies resulting from an application of the Smoluchowski mechanism to a number of experimental data. In particular, it is hard to explain the high activation energies (up to 300 $kJ \cdot mol^{-1}$) observed in the case of recrystallization of metallic catalysts. It is equally difficult to explain why the dimensions of the growing metal crystallites should exceed those of the particles of the porous support. According to these authors, the dominant mechanism for the growth of metallic particles in a supported catalyst is usually Ostwald ripening. The small number of experimental observations of translational movement of small metallic particles are considered to be special cases which do not occur for small metal concentration and larger particles.

In the model of Flynn and Wanke the process of growth of metal particles on the substrate consists of three stages.

1. *Detachment (dissociation) of atoms from small particles*

This step requires the highest activation energy, which is equal to the adsorption energy, E_a of the metal atom on a metallic surface, but is less than the sublimation energy. This energy is of the order 400 $kJ \cdot mol^{-1}$ or less and is easily available at temperatures of 800–1000 K. An oxidizing atmosphere, by increasing the adhesion of metal to the substrate (see above), can also facilitate removal of atoms from a crystallite and transfer to the substrate. Similarly, the presence of defects on the surface of the latter by producing local potential minima, give the same effect.

2. *Migration of atoms on the substrate*

Migrating atoms form a two-dimensional gas, in which the mean translational velocity of the atoms $\bar{v}$ is equal to

$$\bar{v} = \left(\frac{\pi k T}{2m}\right)^{1/2}, \tag{5.27}$$

where m is the atomic mass. This velocity can also be expressed as

$$\bar{v} = a\omega \exp\left(-\frac{E_{sd}}{RT}\right), \tag{5.28}$$

where a is a constant, ω is the vibrational frequency of the surface atoms $\approx 10^{13}\ s^{-1}$ and E_{sd} is the activation energy of surface diffusion (less than E_a).

3. *Capture of the migrating atoms by particles*

The rate of increase of the number of atoms in the capturing particle is

$$\frac{d\nu_z}{dt} = \alpha \frac{\bar{v}\nu_w}{\nu_c S_0} l_i , \tag{5.29}$$

where α is the "sticking coefficient" of the atom colliding with the particle, assumed to be independent of l_i, the effective particle size, ν_z is the number of atoms captured by a particle, ν_w is the number of atoms migrating on the surface, ν_c is the number of atoms on the substrate, and S_0 is the area of the support per metal atom. The total surface area of the support is thus equal to $\nu_c S_0$ and $\nu_w/\nu_c S_0$ is the surface concentration of the migrating atoms.

The atom captured on the periphery of the particle can migrate across its surface. The rate of self-diffusion of the metal is generally significant at normal temperatures, since the activation energy of this process is not too high (for instance, for platinum it is about 110 kJ·g atom^{-1}, which is in the upper range of the values for metals).

The number of atoms in all particles changes with time. The rate of change for the i-th particle is

$$\frac{d\nu_i}{dt} = \frac{d\nu_z}{dt} - \frac{d\nu_u}{dt} = \alpha \frac{\bar{v}\nu_w}{\nu_c S_0} l_i - A\exp(-E_a/RT) , \tag{5.30}$$

where the first term corresponds to the increase of the number of atoms, and the second to its decrease, ν_u is the number of atoms detached from a particle, and A is a preexponential factor.

The number of atoms migrating across the surface should also change during the process, the rate of change being

$$\frac{d\nu_w}{dt} = \sum_{i=1}^{M} \frac{d\nu_u}{dt} - \frac{d\nu_z}{dt} = -MA\exp\left(-\frac{E_a}{RT}\right) + \alpha\bar{v}\frac{\nu_w}{\nu_c S_0}\sum_{i=1}^{M} l_i , \tag{5.31}$$

where M is the number of particles on the support surface area $\nu_c S_0$. Equation (5.31) is entirely analogous to that describing the process of evaporation and condensation between liquid droplets of different sizes.

It follows from equation (5.30) that for certain values of l_i, such that the first and second terms are equal, the net growth of the particle becomes zero. This constitutes another "critical" value of the average effective size of the particle (one has to distinguish it from the critical size, mentioned earlier, above which growth of particle becomes possible on thermodynamical grounds) and is given by

$$l_{i\,\mathrm{crit}} = \frac{A v_c S_0}{\alpha \bar{v} v_w} \exp\left(-\frac{E_a}{RT}\right) = \sum_{i=1}^{M} \frac{l_i}{M} . \tag{5.32}$$

In the course of recrystallization the quantities M and v_w decrease resulting in an increase in l_i. One of the consequences is, that particles, which have grown in the initial stages of the process, decrease in size during the later stages, and $l_{i\,\mathrm{crit}}$, equal to average particle size, increases. In contrast to the Smoluchowski model, the initial size distribution strongly influences the size of particles after recrystallization in the Ostwald mechanism.

The other parameters occurring in equations (5.30)–(5.32) also control the course of the process. The mobility of atoms on the surface depends on the parameter α/S_0, which controls the capture rate and consequently v_w, as well as the translation velocity $\bar{v}$. For large values of α/S_0 and $\bar{v}$, v_w quickly approaches zero and the rate of particle growth is determined by the rate of dissociation. Low values of the parameter α/S_0 can occur either at low metal concentrations (S_0 large) or as a result of strong interactions with the substrate (high value of E_{sd}). In both these cases, accumulation of atoms on the substrate occurs during recrystallization which causes an increase of dispersion during heating (redispersion). This latter process however, is most strongly influenced (exponentially dependent) on the factor E_a/RT and apparently small changes of E_a (for istance by 20 $\mathrm{kJ \cdot g\ atom^{-1}}$) at a given temperature can bring about very significant differences in the rate of redispersion.

The width of the particle size distribution function increases during Ostwald ripening since small particles shrink and the larger ones increase in size, as has often been observed experimentally (see for instance [53]).

The model of Flynn and Wanke gives values of the exponents in equation (5.21) in the range of 2–13 and provides an explanation of changes in these exponents during recrystallization. The values of the exponent are also determined by the breadth of the particle size distribution and the initial sizes of the particles.

Recrystallization of metal particles on a support has been observed for particle sizes differing by at least two orders of magnitude (1–100 nm).

There are reasons to believe that this process cannot be described by a single mechanism. It seems probable that at low initial particle sizes (1–5 nm), crystallite diffusion is the preferred mechanism, while with larger particle sizes, Ostwald ripening may dominate. Over certain temperature and particle size ranges, both mechanisms may operate simultaneously. However, because of the scarcity of experimental data, our knowledge of the recrystallization and redispersion is far from complete.

5.7 INTERACTION BETWEEN THE METAL AND SUBSTRATE DURING THE NUCLEATION OF METALLIC PARTICLES

The nucleation of a solid phase, i.e. formation of the smallest fragments from separate atoms, molecules, or ions is, for obvious reasons, hardly accessible to direct observation. Nucleation occurs both in gaseous and liquid phases, as well as on gas–liquid, liquid–solid and gas–solid phase boundaries and also in the interior of a solid phase (e.g. precipitation of a new phase from a solid solution). The formation of nuclei in the single phase is called *homogeneous nucleation* while that involving a phase boundary is *heterogeneous nucleation.*

Nucleation of metallic phases can be homogeneous, for instance during the evaporation of metals in an atmosphere of an inert gas (see Chapter 2), or precipitation of a solid from a supersaturated solution. Heterogeneous nucleation occurs during the condensation of a metal vapour onto a substrate, or during the reduction of metal compounds to the metal by gaseous or liquid reducing agents.

Formation of nuclei of a metallic phase from a vapour on a solid substrate, is largely governed by the properties of the substrate. The nucleation rate — other conditions, such as vapour pressure and temperature, being equal — depends strongly on the strength of the interaction between the metal and the substrate.

5.7.1 Behaviour of atoms on the substrate

A beam of atoms produced by thermal evaporation of metal in vacuum, impinges onto a substrate at a much lower temperature and condenses. In the immediate vicinity of the surface of the substrate, atoms from the beam are attracted by dipole and multipole interactions and become physically adsorbed. Some time later, the energy of the atoms equals that of substrate by the loss of the kinetic energy component perpendicular to the surface. The atoms then migrate on the substrate surface with random jumps visiting a number of adsorption sites and eventually either desorb back into gaseous phase, or are attached to other atoms forming aggregates. Condensation is thus possible only when the adsorption rate is higher than that of desorption.

Adsorption process may be quantitatively characterized by several parameters; firstly, the *sticking coefficient* expressing the probability that an incoming atom remains attached to the surface and eventually forms an aggregate. It is measured as the ratio of the number of incident atoms to the number condensed on the substrate.

The *thermal accomodation coefficient*, α_T, describes the extent of temperature equilibration of the atom colliding with the substrate surface at temperature T_s

$$\alpha_T = \lim_{T_g \to T_s} \frac{T_R - T}{T_s - T_g}, \tag{5.33}$$

where T_R is the average temperature of atoms reevaporating from the surface and T_g is the temperature of atoms incident on the surface.

The *mean relaxation time*, τ_e, is the time required for equilibration of the temperature of the incoming atoms with the substrate. It is estimated to be less than $2/\omega$, where ω is the frequency of vibration of the bond between the adsorbed atom and the substrate. Lennard-Jones showed that after one vibrational period the potential energy involved in such bonding has fallen to a quarter of its initial value [54].

The *mean residence time* on the substrate, τ_s, is the time between the collision of the atom with the substrate and its desorption

$$\tau_s = \frac{1}{\omega} \exp(Q_{ads}/RT) \tag{5.34}$$

and thus

$$\tau_e = 2\tau_s \exp(-Q_{des}/RT), \tag{5.35}$$

where Q_{des} is the heat of desorption. If the binding energy of the atom to the substrate is sufficiently large, $Q_{des} \gg RT$, and τ_s is very long, and because τ_e is very short, thermal equilibrium is reached very quickly. If however, Q_{des} is of the order of RT, the incoming atoms do not reach thermal equilibrium quickly and can leave the surface, so that the sticking coefficient value becomes much less than unity.

The adsorbed atoms in general, behave like a two-dimensional gas. They move on the surface by random jumps from one adsorption site to another. In the adsorbed state they cover an average distance $\bar{x}$ in time τ_s where

$$\bar{x} = \sqrt{2D_s\tau_s} = \sqrt{2\omega\tau_s}a\exp\left(-\frac{Q_{ads}}{2kT}\right) = \sqrt{2}a\exp\frac{(Q_{des}-Q_{sd})}{2kT}. \quad (5.36)$$

Here, a is the distance covered in one jump, Q_{sd} is the activation energy of surface diffusion (the height of the barrier overcome in changing the adsorption site), and D_s is the surface diffusion constant, equal to $D_s = a^2\exp(Q_{sd}/RT)$. The exact relation between Q_{des} and Q_{sd} is unknown (if there is a general relation at all); it is however generally assumed as a rule of thumb that $Q_{sd} \approx 0.25\,Q_{des}$.

5.7.2 Formation of nuclei

When two atoms moving on the surface collide, they can, at least for a time, form a pair. If dissociation of this pair does not occur first, a third atom may be added to form a trimer and in further steps aggregates with 4, 5, ... atoms will be formed, thus leading to nucleation of the condensed phase.

The rate of nucleus formation depends on τ_s, on the rate of arrival of atoms onto the surface, q, on the rate of adding new atoms to the aggregate λ_i, and on the rate of dissociation of the aggregates, $\varkappa_i$. As a first approximation it can be assumed that at the condensation temperature only single atoms move about and aggregates remain fixed. In this case, the rate of addition of atoms to the aggregates is

$$\lambda_i = w_i n_1(t), \quad (5.37)$$

where $n_1(t)$ denotes the time-dependent concentration of single atoms, $w_i = \sigma_i D_s$ and σ_i is a dimensionless coefficient dependent on the size of the nucleus (but usually taken to be a constant). The rate of formation of aggregates can be expressed more explicitly, using eqs. (5.37) and (5.36) as

$$\lambda_i = w_i n_1(t) = \sigma_1 D_s N_1 \frac{n_1(t)}{N_1} = q\bar{x}^2\sigma_i\frac{N_1}{n_1}, \quad (5.38)$$

where $N_1 = q\tau_s$ is the equilibrium concentration of adsorbed atoms. The quantities $q\bar{x}^2$ and $q\tau_s\bar{x}^2 = N_1\bar{x}^2$ are characteristic parameters of condensation and can be estimated from experimental data. The values of q are normally within the range 10^{12}–10^{14} cm$^{-2}\cdot$s^{-1}, corresponding to 10^{-3} to 10^{-1} monolayers per second, D_s lies within the limits 10^{-6}–10^{-2}, and τ_s can vary over a wide range. For typical experiments, τ may be between 10^{-9} to 1 s. Thus one obtains $\bar{x}^2 = 10^{-14}$–10^{-2}, $q\bar{x}^2 = 10^{-2}$–10^{10} s^{-1} and $N_1\bar{x}^2 = 10^{-11}$–10^{10}.

Aggregates containing only few atoms are unstable and can decompose under certain conditions. The decomposition of larger aggregates can be thought of as evaporation of atoms and, applying the Gibbs–Thomson formula for the vapour pressure over small droplets (see eq. (5.22)), the decomposition rate of the small aggregates can be expressed as

$$\varkappa_i = \frac{1}{\tau_i} = \frac{1}{\tau_0}\exp(-E_i/kT)\,, \tag{5.39}$$

where E_i is the binding energy of the most weakly bound atom in the aggregate, τ_i is an average lifetime of an aggregate composed of i atoms, and $\tau_0 = 1/\omega$ (see eq. (5.34)). For $i > 2$ the value of E_i is normally unknown and thus the rate $\varkappa_i$ cannot be calculated exactly.

It has been observed however [55] that under conditions of very high supersaturation — for example, condensation of the metal from a vapour stream obtained by thermal evaporation onto a substrate at moderate temperatures, giving a ratio of the actual to equilibrium vapour pressures greater than, say, 10^5 — the decay of aggregates larger than 5 atoms is negligible. Thus $\lambda_i \gg \varkappa_i$ for $i > 5$, and nuclei composed of five and more atoms, are therefore, *critical*, that is they do not decompose, but grow. Zinsmeister [56] has even suggested that, in the formation of thin metal films, $\varkappa_i$ can often be neglected for $i \geqslant 2$ so that in such cases the critical nucleus is a single atom. According to this author, $\varkappa_2$ can be estimated from the energy of dissociation of atom pairs E_2. This energy can be estimated, in cases when it is not directly obtainable, as being equal to $\frac{1}{6}Q_{\text{subl}}$ (Q_{subl} is the heat of sublimation). For metals such as Ag, Au, Cu, which condense readily on substrates, $E_2 > \frac{1}{6}Q_{\text{subl}}$, while for those metals which do not condense so easily (their critical nuclei are greater), $E_2 \leqslant \frac{1}{6}Q_{\text{subl}}$.

According to the classical theory of nucleation [57], the size of the critical nucleus is determined by the minimization of the free energy, ΔG, of formation of a spherical nucleus

$$\Delta G = 4\pi r^2\sigma_{\text{sv}} + \tfrac{4}{3}\pi r^3 \Delta G_{\text{v}} \tag{5.40}$$

where r is the radius of a spherical particle, σ_{sv} is the surface tension on the boundary of the vapour and the condensed phase and ΔG_{v} is the difference between the free energy per unit volume of the condensed and vapour phases

$$\Delta G_v = -\frac{RT}{V}\ln\frac{p}{p_e}. \tag{5.41}$$

Here, V is the atomic volume of the condensed phase, p is the actual pressure and p_e is the equilibrium pressure.

A plot of G against r shows a maximum for the critical radius $r = r^*$

$$r^* = -\frac{2\sigma_{sv}}{\Delta G} = \frac{2\sigma_{sv}V}{kT\ln p/p_e}. \tag{5.42}$$

For $r > r^*$, ΔG decreases, hence at higher values of r the particle can only grow, and below this value the particle shrinks. The value of r^* estimated according to this theory is often a fraction of one nanometre, corresponding to clusters of a few atoms and sometimes to single atoms. There is therefore some doubt, whether the application of macroscopic thermodynamic quantities to such small particles is really warranted It seems far more appropriate to consider the phenomenon of nucleation on the basis of quantities definable on an atomic scale.

On the atomic scale, condensation rate can be expressed in terms of the rate of formation of nuclei containing two, three and more atoms. These rates can be expressed by the following differential equations:

$$\frac{dn_1}{dt} = q - \frac{n_1}{\tau_s} - n_1\sum_1 w_i n_i + \sum_2 \varkappa_i n_i \quad \text{(single atoms)},$$

$$\frac{dn_2}{dt} = \lambda_1 n_1 - (\lambda_2 + \varkappa_2)n_2 + \varkappa_3 n_3 \quad \text{(pairs)}, \tag{5.43}$$

. . . .

$$\frac{dn_i}{dt} = \lambda_{i-1}n_{i-1} - (\lambda_i + \varkappa_i)n_i + \varkappa_{i+1}n_{i+1} \quad \text{(clusters of } i \text{ atoms)}.$$

Here it is assumed that the growth of clusters takes place by addition, and decomposition by subtraction of single atoms. In the first equation, n_1/τ_s is the rate of desorption of single atoms from the substrate, $n_i\sum_1 w_i n_i$ is the rate of addition of single atoms to nuclei composed of i atoms and $\sum_2 \varkappa_i n_i$ is the rate of decomposition of nuclei composed of two or more atoms into single atoms and nuclei composed of $n-1$ atoms.

Equations (5.43) cannot be solved in their general form and, in practice, the approximation of Walton [56] is used. This assumes a steady state, i.e. $dn_i/dt = 0$ for $i < i^*$, and consequently $n_i = N_i$, the equilibrium concentration of clusters, composed of i atoms. Thus

$$N_i = \frac{\lambda_1 \lambda_2 \ldots \lambda_{i-1}}{\varkappa_2 \varkappa_3 \ldots \varkappa_i} \quad \text{for } i \leqslant i^*. \tag{5.44}$$

Assuming that λ_i does not depend on the cluster size, $\lambda_1 = \lambda = N_1 D_s$ and with $\varkappa$ defined by equation (5.39) one obtains

$$\frac{N_i}{N_0} = \left(\frac{N_1}{N_0}\right)^i \exp(E_i/kT)\,, \tag{5.45}$$

where N_0 is the number of adsorption sites on the substrate.

The condensation theory allows N_1 and N_i to be calculated as functions of time, the arrival rate of atoms onto the substrate, the condensation coefficient, the density of nuclei and the distribution of their sizes. The following approximations and conditions apply:

1. The substrate is isotropic, homogeneous and free from defects, at least over small areas.

2. Only single atoms can move on the substrate — cluster mobility is negligeable.

3. The mobility of atoms is sufficiently large for an atom to cover a distance $\bar{x}$, larger than the diameter of the nuclei, in time τ_s.

4. The growth of nuclei occurs mainly by the capture of single atoms moving on the substrate.

Adsorption of atoms on nuclei directly from the gaseous phase becomes appreciable only for nuclei of diameter a_i larger than x and in this case

$$\lambda_i = \lambda_{sd} + \tfrac{1}{4}\pi a_i^2 q\,, \tag{5.46}$$

where λ_{sd} is the rate of growth by surface diffusion.

The set of equations (5.43) does not take into account the coalescence of nuclei and the anisotropic distribution of atoms arising as a result of the depletion of atoms in the environment of nuclei. With increasing density and size of the nuclei, the probability that neighbouring nuclei come into contact with each other and coalesce increases, despite their immobility. Moreover, each nucleus acts as a centre capturing atoms from its immediate vicinity, and in consequence a zone of diminished concentration of diffusing atoms forms around the nuclei.

Equations (5.43) have been solved with the above mentioned constraints and the additional assumption that $i^* = 1$ (the critical size of the nucleus is one atom) and w_i is constant. Accurate analysis of experimental data indicates that these assumptions are often realistic. The solution gives $n_1(t)$ and the concentration of nuclei $n_i(t)$ for given q and w_i and for various residence times τ_s. The decrease of adatom concentration in the vicinity of nuclei can be included in the calculations [58] under the assumption that all nuclei are of the same size and the distance between them is the same [59, 60]. The results of the solution of the rate equations are given in the paper of Zinsmeister [61]. Lewis [62] and Stowell [59, 60] also include the effect of the coalescence in the formation of nuclei. The review of Stoyanov and Kaschiev [63] provides the most recent discussion of the solutions of equation (5.43) including the necessary approximations consideration of nucleation and mechanisms rates, capture numbers, cluster densities, spatial distributions of the latter, surface migration and coalescence and references to experimental results.

REFERENCES

[1] M. Humenik and W. D. Kingery, *J. Am. Cer. Soc.*, **37**, 18 (1954).
[2] W. D. Kingery, *J. Am. Cer. Soc.*, **37**, 42 (1954).
[3] R. M. Piliar and J. Nutting, *Phil. Mag.*, **16**, 181 (1967).
[4] H. J. de Bruin, A. F. Moodie and C. E. Warble, *J. Material Sci.*, **7**, 909 (1972).
[5] P. Benjamin and C. Weaver, *Proc. Roy. Soc.*, **A254**, 163, 177 (1960); **A261**, 516 (1961); **A274**, 267 (1963).
[6] C. Weaver, *Chem. and Ind.*, **1965**, 370.
[7] P. Benjamin and C. Weaver, *Proc. Roy. Soc.*, **A252**, 418 (1959).
[8] M. P. Borom and J. A. Pask, *J. Am. Cer. Soc.*, **49**, 1 (1966).
[9] D. M. Mattox, *J. Appl. Phys.*, **37**, 3613 (1966).
[10] J. A. Pask and R. M. Fulrath, *J. Am. Cer. Soc.*, **45**, 592 (1962).
[11] D. C. Moore and H. R. Thornton, *J. Res. Natl. Bur. Std.*, **62**, 127 (1959).
[12] J. F. Hamilton, *J. Vac. Sci. Technol.*, **13**, 318 (1976).
[13] J. F. Hamilton and P. C. Logel, *Thin Solid Films*, **16**, 49 (1973); *ibid.*, **23**, 89 (1974).
[14] J. F. Hamilton and P. C. Logel, *J. Catal.*, **29**, 253 (1973).
[15] B. Lewis, *Surface Sci.*, **21**, 289 (1970).
[16] R. C. Baetzold, *Surface Sci.*, **36**, 123 (1972), *J. Sol. State Chem.*, **6**, 352 (1973).
[17] M. G. Mason and R. C. Baetzold, *J. Chem. Phys.*, **64**, 271 (1976).
[18] D. J. C. Yates, L. L. Murell and E. B. Prestridge, *J. Catal.*, **57**, 41 (1979).
[19] E. B. Prestridge and D. J. C. Yates, *Nature*, London **234**, 345 (1971).
[20] Y. F. Chu and E. Ruckenstein, *J. Catal.*, **55**, 281 (1978).
[21] J. J. Chen and E. Ruckenstein, *J. Catal.*, **69**, 254 (1981).
[22] S. J. Tauster, S. C. Fung and R. L. Garten, *J. Am. Chem. Soc.*, **100**, 170 (1978).
[23] R. T. K. Baker, E. B. Prestridge and R. L. Garten, *J. Catal.*, **59**, 293 (1979).
[24] J. S. Smith, P. A. Thrower and M. A. Vannice, *J. Catal.*, **68**, 270 (1981).
[25] B. J. Tatarchuk and J. A. Dumesic, *J. Catal.*, **70**, 308, 323, 335 (1981).
[26] W. Schottky, *Z. Physik*, **113**, 367 (1939).

[27] N. F. Mott, *Proc. Roy. Soc.*, **A171**, 27, 281 (1939).
[28] G. M. Schwab, J. Block and D. Schultze, *Angew. Chem.*, **71**, 101 (1959).
[29] G. M. Schwab, *Surface Sci.*, **13**, 198 (1969).
[30] R. F. Baddour and M. C. Deibert, *J. Phys. Chem.*, **70**, 2173 (1966).
[31] C. S. Nicolau and H. G. Thom, *Z. anorg. allg. Chem.*, **303**, 133 (1960).
[32] S. M. Skinner, R. L. Savage and J. E. Rutzer, *J. Appl. Phys.*, **24**, 438 (1953); **25**, 1055 (1954).
[33] C. West, *J. Appl. Phys.*, **25**, 1054 (1954).
[34] B. V. Derjaguin and V. P. Smilga, *J. Appl. Phys.*, **38**, 4609 (1967).
[35] W. B. Phillips, E. A. Desloge and J. G. Skofronick, *J. Appl. Phys.*, **39**, 3210 (1968).
[36] J. M. Thomas and P. L. Walker, *J. Chem. Phys.*, **41**, 587 (1964).
[37] M. V. Smoluchowski, *Z. physik. Chem.*, **92**, 129 (1918).
[38] G. M. Fair and R. S. Gemmel, *J. Coll. Sci.*, **19**, 360 (1964).
[39] E. Ruckenstein and B. Pulvermacher, *AICHE Journal*, **19**, 356 (1973); *J. Catal.*, **29**, 224 (1973).
[40] *L.* Wynblatt and N. A. Gjostein, *Progr. in Solid State Chem.*, **91**, 21 (1975).
[41] E. E. Latta and H. P. Bourel, *Phys. Rev. Lett.*, **38**, 839 (1977).
[42] G. Ayrault and G. Ehrlich, *J. Chem. Phys.*, **60**, 281 (1974).
[43] G. Ehrlich und F. G. Hudda, *ibid.*, **44**, 1039 (1966).
[44] E. E. Gruber, *J. Appl. Phys.*, **38**, 243 (1967).
[45] F. A. Nichols and W. W. Mullins, *Trans. AIME*, **223**, 1840 (1965); F. A. Nichols, *J. Appl. Phys.*, **37**, 2805 (1966).
[46] P. Wynblatt and N. A. Gjostein, *Scripta Met.*, **7**, 969 (1973).
[47] J. A. Bett, K. Kinoshita and P. J. Stonehart, *J. Catal.*, **35**, 307 (1974).
[48] W. Ostwald, *Z. physik. Chem.*, **34**, 495 (1900); W. J. Dunning, in *Particle Growth in Suspensions*, Ed. A. L. Smith, Academic Press, London 1973.
[49] B. K. J. Chakraverty, *J. Phys. Chem. Solids*, **28**, 2401 (1967).
[50] C. W. Grenwood, *Acta Met.*, **4**, 243 (1956).
[51] F. H. Huang and Che-Yi Li, *Scripta Met.*, **7**, 1239 (1973).
[52] P. C. Flynn and S. E. Wanke, *J. Catal.*, **34**, 390, 400 (1974).
[53] H. J. Maat and L. Moscou, *Proceed. III Intern. Congress Catal.*, North-Holland Publ., Amsterdam 1965, p. 1277.
[54] J. E. Lennard-Jones, *Proc. Roy. Soc.*, **A163**, 127 (1937).
[55] D. Walton, *J. Chem. Phys.*, **37**, 2182 (1962).
[56] G. Zinsmeister, *Vacuum*, **16**, 529 (1966).
[57] K. L. Chopra, *Thin Film Phenomena*, McGraw-Hill, New York 1969, p. 142; W. Romanowski. *Thin metallic films*, PWN, Warszawa–Wrocław 1974, p. 44 ff (in Polish).
[58] V. Halpern, *J. Appl. Phys.*, **40**, 4627 (1969).
[59] M. J. Stowell and K. J. Rutledge, *Thin Solid Films*, **6**, 407 (1970).
[60] M. J. Stowell and T. E. Hutchinson, *Thin Solid Films*, **8**, 41, 411 (1971).
[61] G. Zinsmeister, *Vakuumtechnik*, **22**, 85 (1973).
[62] B. Lewis, *Surface Sci.*, **21**, 273 (1970).
[63] F. Stoyanov and D. Kashchiev, in *Current Topics in Materials Science*, Ed. E. Kaldis, North-Holland Publ. Comp., Amsterdam 1981, Vol. 7.

6

Interaction Between Gases and Highly Dispersed Metals. Adsorption and Catalysis

6.1 HIGHLY DISPERSED METALS AS ADSORBENTS

Metals, like other solids, act as adsorbents for gases, that is, they retain and condense atoms and molecules incident on a surface. Interactions may be by physical adsorption (physisorption) — with comparatively weak interaction between a gas molecule and the surface metal atoms, or by chemisorption (chemical adsorption), where the interactions are stronger and are due to the chemical affinity between gas and metal. Physical adsorption is a universal phenomenon, occurring on all solid–gas surfaces, the concentration of physisorbed molecules being highest at low temperature. Chemical adsorption, involving strong interactions between the electrons of the solid and of the adsorbate is more specific (selective), producing larger heat effects (both kinds of adsorption are exothermic processes) and usually requires an activation energy with values close to those of chemical reactions involving bulk solids, liquids and gases.

Some of the phenomena associated with adsorption have already been mentioned in the preceding chapter in connection with the adsorption of metal vapours on the surface of solids. The relations between the adsorption energy, residence time, degree of coverage, etc. and the behaviour of adsorbed molecules and atoms are, of course, of more general significance and apply also in the case of adsorption of gases on metals.

It should also be mentioned that the boundary between physical and chemical adsorption is not really sharp. Often there is a doubt whether physical, or chemical adsorption is involved due to the lack of precise diagnostic criteria. The distinction is therefore often only based on convention. Some authors use the energy (heat) of adsorption as a criterion and arbitrarily chose a value of 0.5 eV as an upper limit of the energy of physical adsorption [1].

Most metals and alloys, because of their high chemical activity, interact strongly at moderate temperatures with all except the noble gases. Chemisorption may represent an initial stage of the formation of chemical compounds in the bulk phase, which is controlled by the rate of diffusion of the gaseous reactant through the bulk metal. At lower temperatures, where the diffusion rate is small compared with adsorption, the reaction commonly involves the participation of the uppermost layer of atoms only, i.e. chemisorption occurs. When the reaction goes beyond the first layer of metal atoms and a thin surface film of a new chemical compound is formed, with a thickness of the order of a nanometre to a few micrometres, the phenomena may be termed passivation or corrosion. Finally, at temperatures sufficiently high for diffusion to proceed at a measurable rate, the reaction involves the whole of the volume of the metal.

High dispersion of the metal, by increasing the relative number of surface atoms, enhances chemisorption and reactions with gases become more pronounced. The rate of reaction may be increased as a result of evolution of the heat of reaction which may cause a pronounced rise in temperature. In such cases the reaction may not be limited to chemisorption and violent reactions such as spontaneous combustion of dispersed metals in oxygen or chlorine may result.

The amount of chemisorbed active gases such as H_2, O_2 or CO may be used as a quantitative measure of the surface area of a dispersed metal both as the pure metal and on a non-metallic support. This method had been universally applied over the last two decades, despite some uncertainties concerning the number of molecules reacting with one surface metal atom.

In typical dispersed metal systems, the specific surface area may lie within wide limits: powders 1–20 $m^2 \cdot g^{-1}$, supported metal catalysts up to several hundred $m^2 \cdot g^{-1}$, thin solid films deposited on glass or ceramic up to 100 $m^2 \cdot g^{-1}$. Materials with such large values of the specific surface area make them well-suited for investigations of adsorption phenomena. On the other hand, it may be difficult to obtain high surface purity compared, for instance with samples of bulk metals. Conventional surface cleaning methods suitable for bulk specimens — high temperature heating or ion bombardment — are not feasible for dispersed metals without altering the particle size distribution.

Thus it can be assumed that the purity of the surface of a highly dispersed metal before adsorption is much less adequately characterized than that of a bulk metal and contains many physical and structural defects.

In order to understand the basic features of adsorption of gases on metals it is necessary to look at some of the new methods for investigating the structure and composition of metal surfaces.

6.1.1 Experimental techniques for investigating adsorption on metals

Investigations of the structure and composition of the uppermost atomic layers of solids, and especially metals, have leaned heavily on methods based on the interaction of low energy electrons with the surface. Such investigations are mainly performed on low index surfaces of single crystals, carefully cleaned in ultrahigh vacuum. Both elastic scattering (diffraction) of low energy electrons and inelastic scattering may be used. Furthermore, methods involving the emission of electrons from metallic tips in high, inhomogeneous electric fields — field- and ion-emission — are frequently applied to obtain information on the structure of metallic surface and adsorbed layers. Emission of electrons excited by UV or X-rays (Photoelectron Spectroscopy (PES) or ESCA (Electron Spectroscopy for Chemical Analysis)) can be used to define the type of bonding in the surface layers of the adsorbate and of adsorbent. Measurements of the work function of electrons and changes resulting from adsorption can also give valuable information on the concentration of the adsorbate on the surface. In order to determine the amount of an adsorbed gas and the state of the adsorbing substrate, thermal or electron beam-induced desorption may be useful. In this latter method, the adsorbate–adsorbent system initially in equilibrium at low temperature releases the gaseous adsorbate on heating or electron beam excitation and the resulting pressure increase is monitored by appropriate devices.

6.1.1.1 Low energy electron diffraction (LEED)

A beam of electrons with a kinetic energy of a few tens to a few hundred electron-volts, incident on the surface of a crystalline solid may be elastically scattered with constructive interference. The direction and intensity of the diffracted beams depends upon the geometry of the scattering centres — the atoms of the crystal lattice. This phenomenon, predicted by the wave theory of electrons, was first observed experimentally by Davisson and Germer in 1927 [2]. Electron diffraction by the crystalline lattice is analogous to X-ray diffraction and the Bragg equation may therefore be applied

$$n\lambda = d_{hk} \sin\varphi \,, \tag{6.1}$$

where φ is the scattering angle with an incident beam (usually) normal to the surface, d_{hk} is the distance between the lattice planes on the surface with indices h, k, n is the order of the diffraction, and λ is the electron wavelength. The wavelength λ in nm may be obtained from the electron energy, U in volts,

$$\lambda = \sqrt{\frac{150}{U}} \,. \tag{6.2}$$

The condition for constructive interference at normal incidence is thus

$$\sin\varphi = \frac{1}{d_{hk}}\sqrt{\frac{150}{U}} \,. \tag{6.3}$$

From equation (6.3) it is clear that maxima of diffracted beams occur at the shorter distances, when d_{hk} and/or U increases. The regular array of spots corresponding to the diffracted beams in a plane parallel to the diffracting surface may be observed on a fluorescent screen and constitutes the image of the reciprocal lattice. At accelerating potentials up to a few hundred volts, diffraction occurs almost exclusively at the uppermost atomic layer — scattering from the second and higher atomic planes usually being very weak.

The apparatus for obtaining the low energy electron diffraction images was developed in its present form in the early sixties [3–5]. The basic components are an electron gun producing an electron beam of appropriate energy, a manipulator — a specimen holder which allows translation and rotation of the sample in three directions, with facilities for heating and cooling the sample. The detector consists of at least three spherical wire grids with suitable applied voltages which allow inelastically scattered electrons to be filtered off and elastically scattered electrons to be accelerated thus producing a luminous spot on the fluorescent screen placed behind the grids (Fig. 6.1).

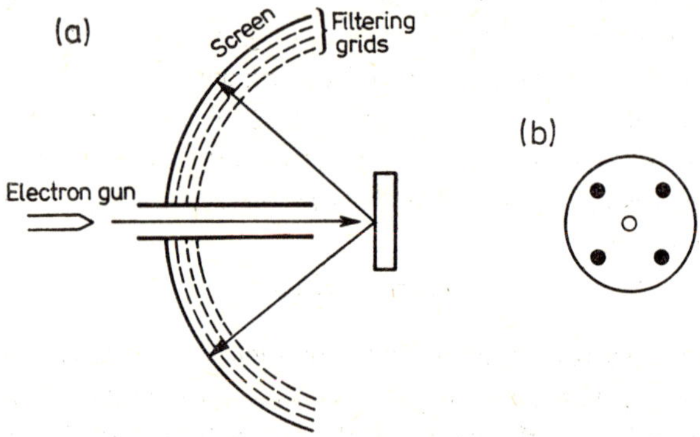

Fig. 6.1 — Schematic representation of low energy electron diffraction (LEED) arrangement (a) and a simple diffraction pattern (b).

The whole diffraction assembly is placed inside an ultra-high vacuum chamber (with a vacuum of the order of 10^{-8} Pa). The specimen is first cleaned by heating to a high temperature, and additionally by bombardment with ions of relatively high energy (several hundred eV) followed by thermal annealing. Occasionally metallic samples may be cleaned by alternate cycles of chemical oxidation and reduction. In the vacuum envelope there are also valves for adding small amounts of the adsorbate gas.

A complete interpretation of the diffraction images involves kinematic and dynamical diffraction theory as well as two-dimensional crystallography, which cannot be described here, even in outline. Details are given in special monographs [6–8], together with a more explicit description of the experimental technique. Nevertheless, it may be useful to point out some of the important results obtained by LEED on the structure of metallic surfaces and adsorbed layers.

The structure of the clean metal surfaces does not always correspond to that of the three-dimensional (bulk) phase. In many cases two-dimensional elementary cells of the surface have different dimensions and symmetry as a result of *reconstruction* of the atomic planes forming the surface. This has been established for elementary semiconductors — silicon and germanium [9], and for metals like platinum [10], gold [11], iridium [12], bismuth and antimony [13]. The reasons for surface reconstruction, seem not to be clear as yet. In the case of germanium and silicon it seems most likely that the free surface orbitals overlap to form new bonds and in so doing minimize the surface free energy. According to some authors, reconstruction of metal surfaces is due to the presence of chemically-active gases such as oxygen or carbon monoxide which cannot be completely eliminated from experimental systems, even under the best vacuum conditions.

The structures of ordered adsorbed layers have been also investigated by LEED. Such ordered layers are formed during both physical and chemical adsorption. The structure — that is the dimensions and symmetry of the unit cells — is determined both by the interaction of the adsorbate with the substrate and by the mutual interaction of admolecules. The structures of adlayers are divided into those characteristic of small adsorbed molecules such as O_2 or CO and those formed by large admolecules such as the heavy noble gases and hydrocarbons. Starting from these notions, attempts have been made to relate the structure of underlying metallic planes to the structure of the adsorbed layer [14]. The LEED method has also been successfully applied to investigations of the adsorption of metallic vapours onto single crystal metallic substrates. This is of great importance in the formation of epitaxial metal layers on metallic substrates.

6.1.1.2 *Investigation of the chemical composition of the surface by means of the Auger electron spectroscopy* (*AES*)

Auger electron spectroscopy may be used to directly determine the chemical composition of the surface of a condensed phase. When excited with a primary electron beam of energy of 2–3 keV, Auger electrons are emitted from the first few surface layers. Every chemical element has a set of Auger transitions with characteristic energies by which the element can be identified. The intensity of the Auger transitions is proportional to the concentration of the corresponding element and this forms the basis of quantitative determinations of the composition. The sensitivity of the analysis is estimated to be about 1% of a monolayer.

The generation and characteristic energies of Auger electrons are shown for the example of silicon, in Fig. 6.2.

A primary electron with an appropriate energy excites an electron from the K-shell of the silicon atom. The hole in the K shell thus formed, is then filled almost instantly by an electron from a higher-lying L-shell, and the energy $E_K - E_{L_i}$ liberated in this process can either be radiated as an X-ray quantum, K, or transferred to other electrons in the L-shell —

say, to the $L_{2,3}$ electrons resulting in one of them being ejected since the difference in energy between the $L_{2,3}$ orbital and the zero, vacuum, level is less than $E_K - E_{L_1}$. The outgoing is an Auger electron with kinetic energy

$$E_k(Z) = E_K(Z) - E_{L_1}(Z) - E_{L_{2,3}}(Z+\Delta) - \varphi \,, \tag{6.4}$$

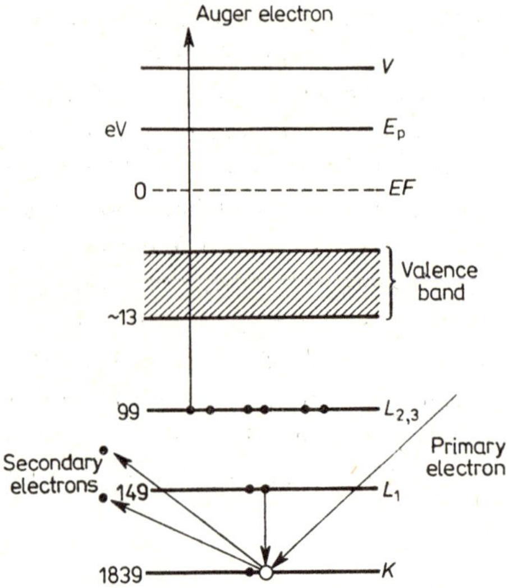

Fig. 6.2 — Auger electron generation in silicon (after [15]).

where Z in brackets indicates that the corresponding energies are dependent on the atomic number, Z, of the element, and φ is the work function of the solid element. The whole process is denoted as a $K L_1 L_{2,3}$ Auger transition. In this process the atom becomes doubly ionized and since after ejection of the K-shell electron, the atom is ionized, a correction term, Δ, is included in equaticn (6.4) where $\Delta \approx 1$.

Relaxation of the atom after the first ionization process thus involves either the emission of a fluorescent X-ray quantum, or of an Auger electron. These two processes have different probabilities depending on the atomic number, Z, as shown in Fig. 6.3.

For elements with small atomic number ($Z < 11$), the probability of an Auger transition is near unity and decreases to about 0.5 at $Z \simeq 33$. Auger spectroscopy is thus particularly sensitive for the lighter elements present on the surface. Detection and accurate intensity measurements are not simple matters, since the Auger electrons are emitted from the specimen surface together with other secondary electrons. The corresponding $N(E)$ — curve (number of secondary electrons versus their energy) shows a peak near zero energy due to secondary, multiply scattered electrons, and another peak close to E_0 due to elastically scattered electrons. These contribute a high noise level with small, barely noticeable peaks, arising from the Auger electrons. In order to improve the recording and intensity measurement of Auger electrons, the derivative of $N(E)$ rather than the $N(E)$ curve itself is recorded.

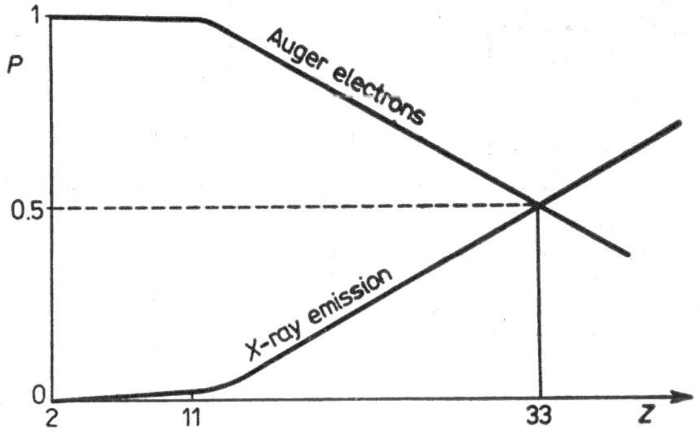

Fig. 6.3 — Relative probabilities, P, of X-ray and Auger emission as a function of atomic number Z (schematic).

The apparatus for Auger electron detection and intensity measurement consists, like that for LEED measurements, of an electron gun producing electrons with variable energy, an appropriate sample holder and a secondary electron energy analyser with a voltage modulation device allowing direct recording of dN/dE versus E. Instead of the four-grid electrostatic electron energy analyser mentioned earlier, a double focussing cylindrical mirror analyser (CMA) is often used, which has the advantage of better energy resolution. The retarding grid field analyser can however be used to perform LEED analysis also thus allowing the possibility of performing LEED and AES measurements in the same apparatus.

6.1.1.3 Photoelectron spectroscopy (PES) and electron spectroscopy for chemical analysis (ESCA) [16]

Surface atoms can be ionized by excitation with X-rays of a suitable wavelength or by ultraviolet radiation. The photoelectrons thus produced have kinetic energies equal to the difference between the energy of the exciting quanta and the energy of the atomic states from which the electrons were excited. The energy balance for the photoemission of an electron from a sample (see Fig. 6.4) is

$$h\nu = E_b + \varphi_{sp} + T_{sp} + E_r \,, \tag{6.5}$$

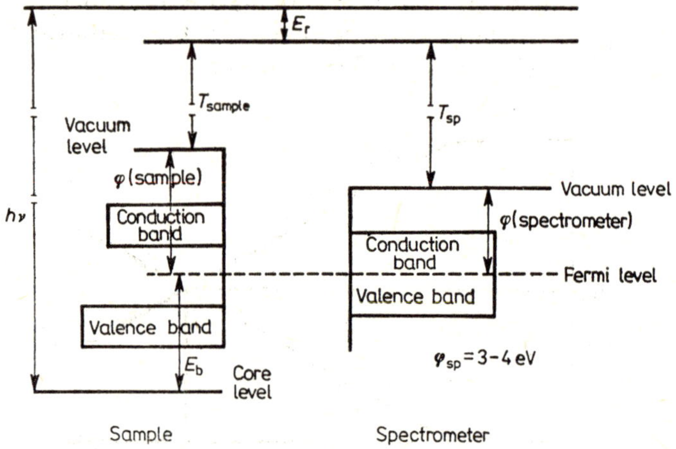

Fig. 6.4 — Inner ionization scheme in the ESCA (PES) method.

where $h\nu$ is the incident photon energy, E_b is the binding energy of an electron from a core state or an outer shell, φ_{sp} is the work function of the spectrometer material, T_{sp} is the kinetic energy of the electron, measured by the spectrometer, E_r is the recoil energy of the emitting atom. In the experiment, $h\nu$ is *a priori* known, T_{sp} is measured, and the quantities φ_{sp} and E_r are characteristics of a given spectrometer and sample and are constant for a given experiment.

Analysis of the photoelectron energy spectrum, as for Auger electrons, allows the atom type and concentration in the surface layers to be measured. Excitation with UV radiation causes ionization of the outer (valence) electron shells, while excitation with X-rays (usually those of *K*-series of magnesium or aluminium) causes the ionization of the inner shells. In both cases the exact value of the photoelectron energy depends on the "chemical environment" — that is on the nature and strength of the bonding with neighbouring atoms. A difference in the binding energy of inner electrons of a given element in different chemical compounds is called a "chemical shift". Measurement of the chemical shift, which can be obtained more accurately using ESCA than AES, makes it possible to determine not only the elements, but also the chemical compound present in a surface layer. The probe depth is similar to that for AES, due to the comparatively small penetration depth of UV radiation and soft X-rays.

With this method it is thus possible to investigate the electronic structure of the uppermost layers of the adsorbents and that of adsorbed layers. A large number of such investigations have been made, for example, the adsorption of oxygen and carbon monoxide on nickel [17] (with a theoretical interpretation by the method of quantum chemistry [18–21]), the adsorption of ethylene, acetylene and benzene on nickel [22], and of oxygen on the (110) and (100) planes of tungsten [23, 25], on polycrystalline nickel [24] and on a (110) surface of silver [25].

6.1.1.4 Field emission of electrons from metals [26]

Tunnelling of electrons through a surface potential barrier lowered by an external electric field (Fig. 6.5), constitutes the basis of the field emission and ion microscopes invented by E. Müller [27].

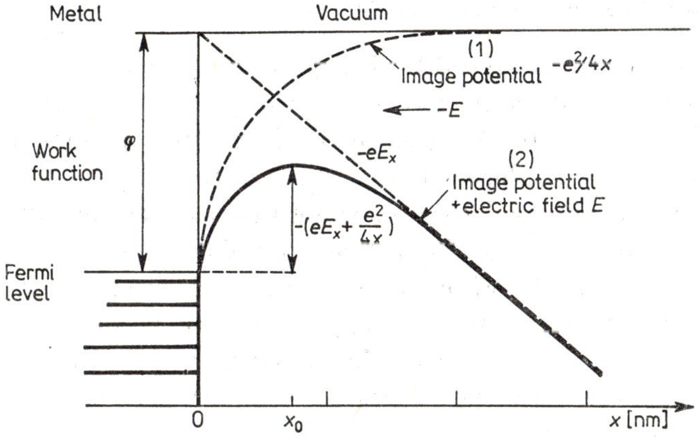

Fig. 6.5 — Potential barrier at the metal–vaccum interface.

In an applied electric field, electrons in the metal in energy levels close to the Fermi level can penetrate the modified potential barrier and move parallel to the electric field. At constant field strength, the intensity of electron emission depends on the position of the Fermi level of the metal, that is, on the work function φ, the latter having different values for different crystal planes. Thus, if a strong electric field is produced in the vicinity of a single crystal of a metal with different crystal planes exposed, the emission image consists of a pattern formed by electrons emitted from the surface with different intensities.

A schematic diagram of a simple *field microscope* which is not too dissimilar from that designed by Müller, is shown in Fig. 6.6.

The electron-emitting surface consists of a very thin, monocrystalline tip of the metal under investigation with a radius of curvature of 50–100 nm.

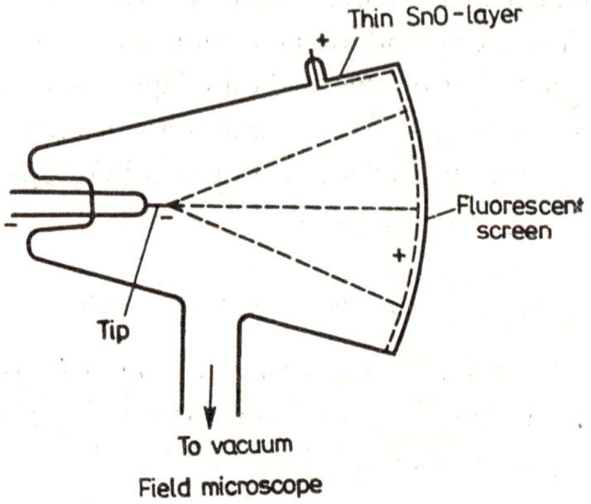

Fig. 6.6 — Field electron microscope (schematic).

A negative voltage of several kV is applied, and the anode is positioned near the fluorescent screen, giving a field of several volts per nanometre in the vicinity of the tip. The small dimensions of the tip in relation to the size of the screen produce a magnification of the image of the order of 10^5. On the screen, light spots of various intensity are observed, corresponding to the local emission current from different regions of the tip. If the crystallographic orientation of the tip is known, it is possible to assign Miller indices to the lattice planes constituting the surface of the tip with the aid of a crystallographic projection and identify the luminous spots as arising from various crystallographic planes forming the surface of the tip (Fig. 6.7).

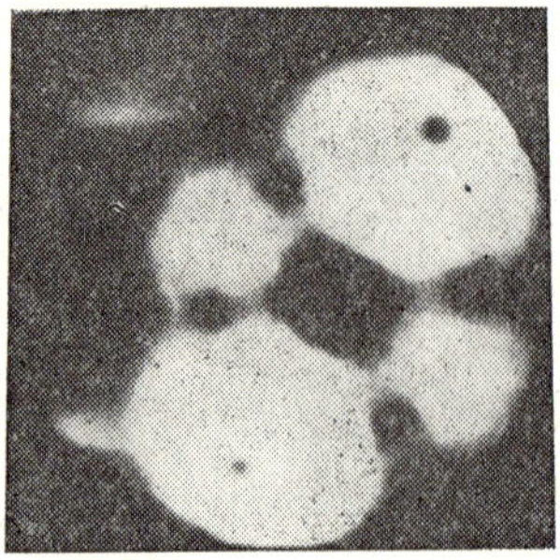

Fig. 6.7 — Field emission picture of the tungsten tip (Figs. 6.7 and 6.8 were taken by Dr A. Ciszewski, Institute of Experimental Physics, Wrocław University, Poland).

Any interaction which influences the magnitude of the work function of the planes constituting the surface produces changes in the intensity of the light spots forming the image. Specifically, any molecules or atoms adsorbed on the surface produce changes in the work function of the adsorbing plane and corresponding changes in emission. It is thus possible to detect adsorption, to obtain information on the concentration of adsorbate on different crystal planes and to follow the diffusion of the adsorbate on single crystal surfaces. Thus qualitative, and sometimes even quantitative conclusions can be drawn about the interaction of the adsorbate with the metal and the anisotropy of adsorption and diffusion.

Still more accurate information about the structure of metal surfaces and adsorption can be obtained with the aid of an *ion microscope* [28–30]. The ion microscope is similar to the field microscope, but the glass bulb surrounding the emitter and screen contains a gas such as H_2, He or Ne at low pressure ($\approx 10^{-1}$ Pa) instead of a high vacuum. A positive voltage is applied to the tip and the screen acts as the cathode. The electric field at the anode is several times higher than that at the cathode of a field ion microscope, and causes the polarization and ionization of hydrogen or other imaging gas. The process of formation of ions at the tip is complicated and not understood in detail at the moment. However, the resulting H^+ ions leaving the ionization centres (surface atoms) have a tangential component of velocity which is much smaller than that of the electrons in a field microscope. Consequently, an image of much greater resolution is obtained and this allows single ionization centres (atoms) to be observed at distances of a fraction of a nanometre (Fig. 6.8). The resolving power δ[nm] of a field ion microscope is approximately equal to

$$\delta \approx \left(\frac{6 \times 10^{-4} Tr}{F}\right)^{1/2}, \tag{6.6}$$

where T is the temperature [K], r is the radius of curvature of the tip and F is the electric field [$V \cdot nm^{-1}$] [28].

In this way, a much more accurate visualization of the surface structure can be obtained and changes due to the influence of temperature, electric field and adsorption of gases other than the image gas, etc.

Field electron and ion microscopes have been used for a number of years as important tools for investigation of the structure of adsorbed layers and for the observation of the kinetics of processes such as surface reconstruction, growth of epitaxial layers and so on. The limitation of these methods lies in the fact that cleaning the tip surface requires flash heating to a high temperature. Sometimes the so-called field desorption

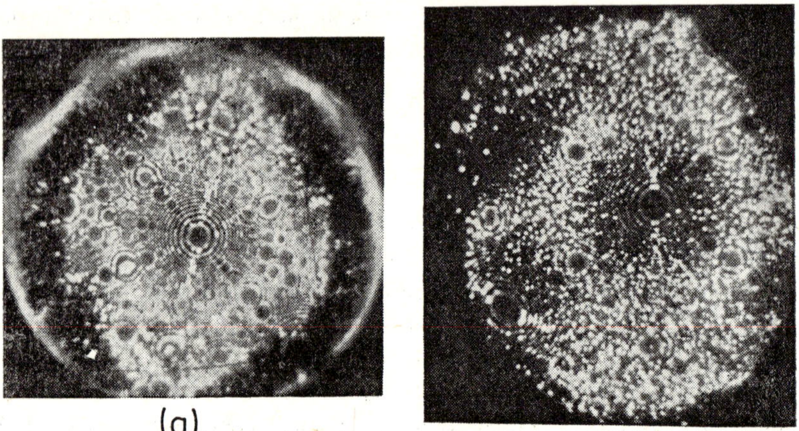

(a) (b)

Fig. 6.8 — Ion emission picture from (a) a clean tungsten tip, (b) the tip with adsorbed residual gases and indium vapours.

of a surface layer of atoms can be used for this purpose also. Consequently only metals with a high melting temperature and cohesive energy are suitable as tip materials. A further limitation is that observations are made in the presence of a high electric field which can alter the behaviour of adsorbate and influence reconstruction processes on the metallic surface. The most valuable advantage of these methods undoubtedly lies in their ability to allow the observation of the structural details of the metal surface and adsorption processes on an atomic scale, at low temperatures and on a substrate which has a well-defined composition and structure.

6.1.1.5 Electron- and thermally-stimulated desorption

In order to determine the quantity and nature of an adsorbate, a desorption technique is often useful. Desorption can be induced by bombardment of the adsorbate layer with electrons, or by heating to high temperatures, the adsorbate appearing in the gas phase as neutral molecules or ions. The desorption probability may be defined as the ratio of the intensity of the desorbed ion or molecule beams to the intensity of the incident electron beam $w = i_{\pi.\mathrm{ol}}/i_{\mathrm{el}}$. Values of $w = 10^{-6}$–10^{-4} are obtained for ions and $w = 10^{-2}$ for neutral molecules, using 100 eV electrons, which have been shown to give maximum desorption yield.

The energy transferred from an incident electron impacting an adsorbed atom (or molecule) is in most cases small compared with the adsorption energy and cannot, therefore, directly induce desorption. A proposed mechanism for desorption is based on the Franck–Condon principle [31]: collision of an electron with an adsorbed atom results in ionization of the atom, without changing the distance between the nuclei of the adsorbate and substrate atoms and their respective velocities. The ionized adsorbed atom is then transferred into a state of higher potential energy which makes the desorption possible without further intake of energy.

The mechanism by which neutral atoms are desorbed as a result of electron excitation was initially unclear; now it is assumed that neutral atoms are created by the excitation of adsorbed atoms to antibonding states. This is preferred to a mechanism which postulates the neutralization of a desorbed ion by the incoming electrons [32].

Electron-beam stimulated desorption of gases like hydrogen and carbon monoxide has been extensively investigated, both for single-crystal and polycrystalline metallic adsorbents [33, 34].

The energy of adsorption is however, more often determined by thermal desorption. In this method, metallic wires or ribbons may be used as adsorbents, which are then rapidly heated electrically at several hundred degrees per second. The adsorbed gases are thus rapidly desorbed and the resulting pressure change is measured with an ionization manometer (assuming a constant pumping rate). Differential pressure versus temperature plots obtained in this way are called *desorption spectra* and show, as a rule, several maxima, and a number of "humps" and "shoulders" resulting from overlap (see Fig. 6.9).

It is possible to apply the thermal desorption technique to thin metal films evaporated in vacuum, and to powders also. In these cases however, a much slower heating rate is used [35]. Prior adsorption of the gas is carried out on a carefully degassed adsorbent, at low pressures (not exceeding 10^{-3} Pa) and desorption is measured after evacuation of the adsorption cell at the lowest possible temperature.

In order to interpret thermal desorption curves it is necessary to know the rate of temperature change during desorption, $T = f(t)$. In most cases heating conditions are chosen so that the temperature is a linear function of time: $T = T_0 + bt$. Then the highest desorption rates, corresponding to the maxima of the curve $p = f(T)$, occur at temperatures $T = T_{max}$, related to the activation energy of desorption, E

$$\frac{E}{RT_{max}^2} = \frac{\nu_1}{b} \exp\left(-\frac{E}{RT_{max}}\right), \tag{6.7}$$

where R is the gas constant, ν_1 ($\approx 10^{12}\,s^{-1}$) is the frequency factor for first order desorption (i.e. when the desorption is not accompanied by the change in the molecular state of the adsorbate). Equation (6.7) can also be expressed in a simplified form [36]

$$E = RT_{max} \ln\left(\frac{\nu_1 T_{max}}{b} - 3.64\right), \tag{6.8}$$

where E is measured in $kJ \cdot mol^{-1}$.

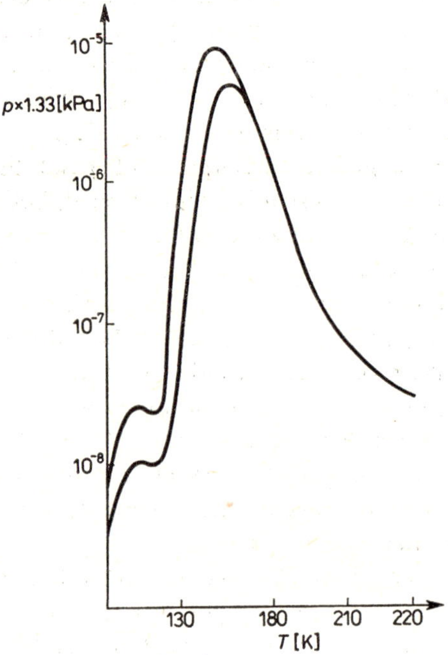

Fig. 6.9 — Desorption spectrum of nitrogen from polycrystalline cobalt (after P. A. Redhead [36]).

For desorption processes of higher order (when desorption proceeds with dissociation or association of desorbed molecules) other, more complicated relations apply. The activation energy of desorption is the sum of the activation energy of adsorption and the heat of adsorption. Since physical adsorption does not require activation, $E = Q_{ads}$ in this case.

The interaction between adsorbate and adsorbent is, in general, not identical at every site on the adsorbent surface. It depends both on the nature of the sites where the adatom or admolecule is bonded — what may be called *a priori* heterogeneity of the surface — and on the concentration of molecules already adsorbed — the *a posteriori* heterogeneity. This is reflected in the adsorption spectrum by the occurrence of many peaks at various values of T_{max} and by the broadening and deformation of the peaks. The different desorption peaks can be interpreted in terms of different *forms of adsorption* of the given adsorbate on a given adsorbent. They may correspond to admolecules, at lower T_{max} values, and to adatoms at higher T_{max} (when adsorption at higher temperatures occurs with simultaneous dissociation). The presence of different types of adsorption is clearly governed by the nature both of the adsorbent and the adsorbate. During heating of the adsorbent–adsorbate system it is also possible to observe a transition from one type of adsorption to another; for instance a transition from the mobile (two-dimensional gas) into a localized one, or the transition of an admolecule from one site to another with a different binding energy, etc. An accurate interpretation of the various forms of adsorption, represented in the desorption spectrum is possible only if sufficient knowledge is available from other methods concerning the type of bonding, polarization of the adsorbate molecule, its electronic structure and other atomic scale structural parameters. Nevertheless, the desorption method is capable of giving a useful overall picture of the energetics of the adsorption process.

Desorption experiments are now normally conducted on polycrystalline adsorbents, such as thin metallic films whose surface is composed of crystallites with different orientations. In these cases, only a slow rise of temperature is practicable during desorption and the temperature range extends from cryogenic temperatures up to 500–600 K — otherwise the surface structure would be altered by prolonged heating. The surface area of such adsorbents is sometimes two orders of magnitude higher than that of wires and ribbons, but to obtain a sufficiently clean surfaces ultra-high vacuum conditions and, possibly, the purest adsorbates are needed.

6.1.1.6 Adsorption-induced changes in the work function

Adsorption on metals results in changes in the electron work function which is defined as the energy $e\varphi$ necessary to transfer an electron from the Fermi level to the exterior of the crystal (the "vacuum level"). On a clean crystal surface there is always an electric dipole layer, originating from the inhomogeneous distribution of electron density in the vicinity of the crystal boundary. This layer is characterized by a so-called surface potential, $\chi = \varphi_0 - \varphi_i$ (see Fig. 6.10), depending on the geometrical ar-

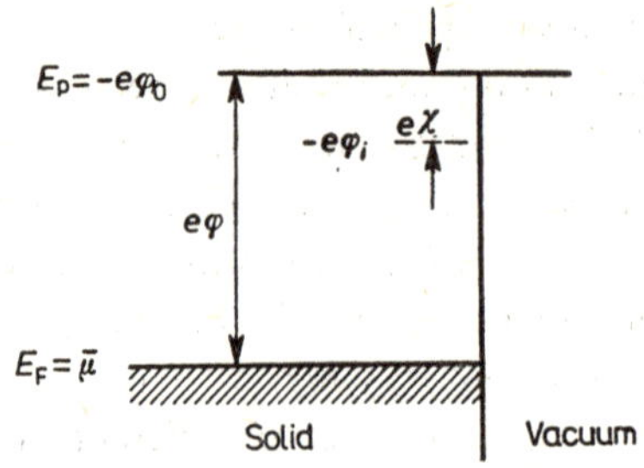

Fig. 6.10 — The surface potential of a clean metal surface.

rangement of atoms on the surface, and thus having different values for different crystal planes constituting the crystal surface. In the case of a polycrystalline surface composed of many different lattice planes, an average potential, $\bar{\chi}$ applies at small distances from the surface and an average value for the work function is obtained.

In the presence of an adsorbed layer a new double layer is formed on the surface and the surface potential changes. The work function increases or decreases, depending on the direction of the adsorbate dipoles, by the same amount. The surface potential after adsorption is defined as

$$\Delta V = \varphi_c - \varphi_a = -\Delta\varphi , \tag{6.9}$$

where φ_c denotes the work function of a clean surface, and φ_a is the work function of the same surface after adsorption.

Measurements of the work function and changes resulting from adsorption can be carried out by several methods since these quantities are included in the quantitative description of many phenomena involving the emission of electrons from the surfaces of solids. Measurement of the thermal emission of electrons and the treatment based on the Richardson–Dushman equation (see Chapter 3) can be used for clean metals, but not for adsorbed layers, since the method involves heating the metal to a high temperature, at which the adsorbate would evaporate or be otherwise destroyed. Another possibility is offered by the photoelectric effect (external photoemission), where illumination with quanta $h\nu > e\varphi$ gives rise to an electron current, i, given by the Fowler equation

$$i = BT^2 f(h\nu - h\nu_0)/kT . \tag{6.10}$$

Here, B is (nearly) a constant in the vicinity of the threshold frequency ν_0 and f is a known function of φ [37]. Both of these methods, if applied to polycrystalline surfaces composed of different crystallographical planes, give measured values of φ which are not equal to the weighted average $\bar{\chi}$, but are preferentially biased towards the lowest local work functions. This is true even if the fraction of the surfaces area corresponding to low φ is small and appreciably lower values of the work function are obtained therefore.

The field emission current (see 6.1.1.4) also depends on the work function of the metallic emitter according to the Fowler–Nordheim equation

$$i = \frac{e^2F^2}{8\pi h\varphi t^2(y)} \exp\left(-\frac{4(2m)^{1/2}\varphi^{3/2}}{3heF}\right) f^*(y), \tag{6.11}$$

where F is the electric field strength [$\mathrm{V \cdot cm^{-1}}$], $y \sim F^{1/2}\varphi^{-1}$, $t(y)$ and $f^*(y)$ are correction factors which are close to unity. Thus a graph of $\log i/F^2$ against $1/F$ should yield a straight line, and from its slope the value of φ may be obtained. This method is often used, firstly to determine φ for individual planes of a single crystal emitter and, secondly, for measurements of changes in the work function induced by the presence of adsorbates. The latter technique is made possible by measurements of the local emission current from one plane of the emitter, with the aid of movable Faraday cage positioned inside the field microscope tube.

The methods outlined above allow a determination of absolute values of the work function. However, for the detection of adsorption and for measurement of the changes in φ due to adsorption, a comparative measurement, of, for example, the contact potentials of two metals, $V_{AB} = \varphi_A - \varphi_B$, may suffice, providing that φ_A or φ_B are known. Such measurements can be carried out with great accuracy by one of the methods described below.

Vibrating capacitor method. When two conductors A and B form the plates of a capacitor with only a small separation between them, potential difference $V_{AB} = \varphi_B - \varphi_A$ corresponds to a charge Q in the capacitor

$$Q = CV_{AB}. \tag{6.12}$$

The capacity C is determined by the geometry of the plates, and can be varied in time by vibration of one of them with a frequency $\omega/2\pi$

$$C(t) = C_0 + C\sin\omega t. \tag{6.13}$$

A current is induced in an external circuit where

$$i(t) = \frac{dQ}{dt} = V_{AB}\frac{dC}{dt} = V_{AB}\omega\Delta C\cos\omega t. \tag{6.14}$$

V_{AB} may be measured by compensating this voltage with an equal external voltage so that when the voltage difference between the capacitor plates is zero, any change of capacity C no longer induces a current, i.e. $i(t) = 0$. When V_{AB} changes as a consequence of adsorption on one of the plates, the compensating voltage also changes and this voltage is recorded automatically.

In this method, the work function of one of the capacitor plates should be independent of adsorption, i.e. the reference electrode should not adsorb the gas in question. Good materials for reference electrodes are gold, oxidized tantalum, or oxidized tungsten. Experimental details of the vibrating capacitor method are described by Craig and Radeka [38].

Diode method. A diode may be formed by an incandescent cathode of a metal with a known work function, such as a tungsten wire, and a thin film of the metal under examination deposited on the walls of the glass bulb surrounding the cathode and forming the anode. The anode cur-

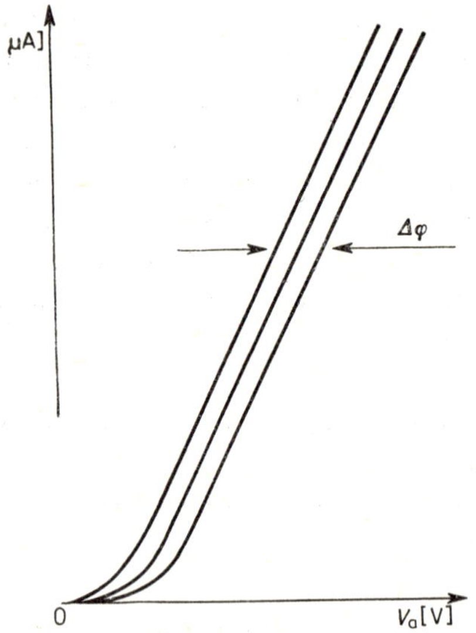

Fig. 6.11 — The current-voltage characteristics of a diode (schematic).

rent i_a of such a diode, working either in the retarding field or space charge limited regimes, depends on the difference between the applied voltage V_a and the work function of the anode, φ_A. A typical series of graphs of i_a vs V_a is shown in Fig. 6.11.

When a gas is adsorbed on the anode, the value of φ_A is changed and the current-voltage characteristic is shifted parallel to the V_a axis by the value $\Delta\varphi_A$. The current-voltage characteristics are automatically recorded. A good review of several variants of this method has been given by Knapp [39].

In all the cases of adsorption investigated by this method to date, the work function of the metal adsorbent increases or decreases as a result of a modification of the surface dipole layer by the adsorbate. When a chemisorptive bond is created, there is an electron transfer between the adsorbate and the metallic substrate, the direction of this transfer depending on the nature of both adsorbate and adsorbent. In the case of physical adsorption, the adsorbed molecules are polarized by the surface electric field. In both cases a dipole layer is formed and the work function change, $\Delta\varphi$ depends on the surface density n_s of adsorbed molecules and their dipole moment, μ_s. Assuming that the adsorbed molecules form a continuous and infinite double layer of charges, and the distance between them is d, the expression for the dipole moment per unit surface area is

$$n_s\mu_s = \varrho d\,, \tag{6.15}$$

where ϱ is the surface charge density. The capacity per unit surface area, $C = \varepsilon_0/d$ is related to the potential change $\Delta\varphi$

$$\Delta\varphi C = \varrho \tag{6.16}$$

and thus

$$\Delta\varphi = n_s\mu_s/\varepsilon_0\,; \qquad \mu_s = \frac{\varepsilon_0}{n_s}\ (\mu_s \text{ in C}\cdot\text{cm}) \tag{6.17}$$

(1 C·cm = 3×10^{27} Debye units, $\varepsilon_0 = 8.85\times10^{-14}$ C·volt^{-1}·cm^{-1}).

Under the assumption that μ_s is independent of the surface coverage θ, $\Delta\varphi$ should change linearly with coverage. In reality however, dipoles present on the surface interact with each other leading to a mutual depolarization and a decrease in the dipole moment. This effect is described by the Topping relation [40]

$$\Delta\varphi = \frac{n_s\mu_s^0}{\varepsilon_0(1+9\alpha n_s^{3/2})}\,, \tag{6.18}$$

where μ_s^0 is the dipole moment of adsorbate at the zero coverage, and α is the polarizability of the adsorbed molecules.

The Topping relation however, has proved to be unsatisfactory for the case of adsorption of low work function metals on other metals, where a minimum in the $\Delta\varphi-\theta$ relation is observed. The interaction of adsorbate dipoles with their electrostatic image in the adsorbent results in a mutual compensation of the change in dipole number and a decrease of the moment at some definite coverage. This case is better described by Gyftopulos–Levine formula [41]

$$\frac{\Delta\varphi}{\varphi_{\text{sub}}-\varphi_{\text{ads}}} = 1-G(\theta)\frac{A}{1+B\theta^{3/2}} \,. \tag{6.19}$$

Here, φ_{sub} and φ_{ads} are the work functions of substrate and adsorbate respectively, A and B are constants, θ is the coverage and $G(\theta) = 1 - 3\theta^2 + 2\theta^3$. The relation between $\Delta\varphi$ and the coverage for alkali metals on transition metals with a large value of φ was worked out theoretically by Lang [42] by a quantum mechanical SCF treatment of a "jelium" model. The results were in good agreement with experiment.

In the case of chemisorption of gases on group IB and transition metals, the situation is even more complicated (than for metal/metal adsorption) since various forms of chemisorption are possible and several types of relationships between $\Delta\varphi$ and θ have been observed for these systems [43–45].

6.1.2 Investigation of adsorption on highly dispersed metals

6.1.2.1 Adsorption equilibria [46]

Highly dispersed metals (powders, thin films, metals dispersed on a carrier) have wide practical application and thus a knowledge of their adsorption properties is of great importance. Differences between highly dispersed and massive metals lie not only in larger specific surface area but also in the fact that adsorption centres of several different kind can occur on the surface of a dispersed metal when comparison is made with a bulk metal and especially with single crystal surfaces.

The concentration of adsorbed gases and vapours per unit mass may thus be orders of magnitude greater for dispersed metals, and this facilitates accurate measurements.

The amount of an adsorbed substance depends on the surface area of the adsorbent, pressure and temperature. To describe equilibrium adsorption conditions quantitatively, the *adsorption isotherm* is frequently used, namely the quantity of adsorbed gas in equilibrium with the gaseous phase as a function of pressure at a given constant temperature. In some cases the dependence of the pressure of a gas in equilibrium with the adsorbed phase with temperature at constant coverage (adsorption isostere) may also be used. A further way to describe equilibrium adsorption is to establish the *adsorption isobar*, i.e. the dependence of the concentration of adsorbate on temperature at a constant adsorbate pressure.

Expressions for the adsorption isotherms can be derived theoretically on the basis of chosen models for adsorption. The most frequently used adsorption isotherms are the following:

1. *The Langmuir isotherm*

$$\theta = \frac{bp}{1+bp}, \tag{6.20}$$

where θ is the coverage, that is, the ratio of the number of adsorbed molecules to that necessary to cover the surface with a monomolecular layer, p is the pressure of adsorbate at equilibrium, and b is the *adsorption constant* — equal to the ratio of rate constants for adsorption and desorption. The model underlying this isotherm assumes adsorption to be complete with a monolayer of adsorbate, and the surface energetically homogeneous. Thus, all adsorption centres have the same adsorption energy, and the adsorbed molecules do not interact with each other. These assumptions are clearly somewhat unrealistic; nevertheless the Langmuir isotherm gives a good representation of many experimental results. In particular, the Langmuir isotherm has proved to be valid for chemisorption of gases on metals, including cases where adsorption proceeds with dissociation of diatomic molecules, when the Langmuir isotherm has the form

$$\theta = \frac{(bp)^{1/2}}{1+(bp)^{1/2}}. \tag{6.20a}$$

2. The *Freundlich isotherm* was originally obtained as an empirical correlation and is expressed by

$$\theta = kp^{1/n}, \tag{6.21}$$

where k and n are temperature-dependent constants. This formula can be also derived theoretically with the assumption that the adsorption energy decreases logarithmically with coverage, and relating, therefore, to a particular form of surface inhomogeneity.

3. The *Temkin isotherm.* When the heat of adsorption, Q, depends linearly on coverage,

$$Q = Q_0(1-\alpha) \tag{6.22}$$

(Q_0 is the heat of adsorption for zero coverage (initial adsorption heat) and α is a constant), the Temkin isotherm is obtained

$$\theta = \frac{RT}{\alpha Q_0} \ln(Bp) , \tag{6.23}$$

where B is a constant.

For physical adsorption of gases on metals, the following isotherms may be used.

4. The *Brunauer, Emmett and Teller (BET) isotherm*

$$\frac{p}{V(p_0-p)} = \frac{1}{V_m C} + \frac{(C-1)p}{V_m C p_0} . \tag{6.24}$$

Here V is the volume (reduced to STP) of the adsorbate at pressure p, p_0 is the saturation pressure of the adsorbate at the temperature of measurement, C is a constant dependent on the heat of adsorption for the first molecular layer and the heat of condensation. This isotherm is derived from a multilayer adsorption model and is used to determine the value V_m — i.e. the STP-volume of adsorbate corresponding to a monolayer coverage. If this volume and the surface area per molecule of adsorbate is known, the surface area of the adsorbent sample can be determined. This isotherm is most often used for the determination of the specific surface areas of all kinds of highly dispersed solids.

To determine the surface area of highly dispersed transition metals, both as pure powders and supported on carriers, the specific chemisorption of reactive gases such as hydrogen, oxygen and carbon monoxide is often used. After reaching saturation coverage at the appropriate temperature (the Langmuir isotherm being appropriate for this case) values of V_{sat} are obtained from which the sample surface area may be calculated. The use of this method however, requires knowledge of the "stoichiometry", i.e. the number of adsorbate atoms (molecules) reacting with one surface atom of metal at saturation coverage.

5. The *Dubinin–Radushkevich isotherm* applies up to a monolayer coverage for energetically heterogeneous surfaces and for small relative pressures

p/p_0 (up to 0.13)

$$\log N_a = \log N_m - \frac{B}{2.303} \varepsilon^2, \tag{6.25}$$

where B is a constant depending upon the average adsorption energy and ε is the so-called Polanyi potential equal to $RT \ln(p/p_0)$. The constant B is also related to the isosteric heat of adsorption

$$Q_{\text{isoster}} = Q_c + \left(\frac{\ln\theta}{B}\right)^{1/2}, \tag{6.26}$$

where Q_c denotes the heat of condensation of the adsorbate and $\theta = N_a/N_m$ is the coverage.

Experimental measurements of adsorption are carried out using appropriate quantities of pure, carefully degassed adsorbents (obtained by degassing by heating in high vacuum). The gas to be adsorbed is admitted in calibrated volumes and the pressure measured after equilibrium is established. In the case of dispersed metals, preparation of samples requires conditions which exclude the presence of highly reactive gases, and degassing should be equally carefully performed, using the highest possible vacuum and avoiding high temperatures; otherwise there is a risk of irreversible reaction between the gases and the adsorbent and a reduced surface area due to sintering. Experiments may be conducted over a very wide pressure range (10^{-1}–10^7 Pa). To measure the pressures accurately, various types of manometers may be used, but they should neither pump the gas nor cause dissociation, and ionization manometers are rarely therefore the best choice for measurements at the lowest pressures. Such problems, although very important experimentally, cannot be considered here in detail and the interested reader is referred to specialist publications [48].

Adsorption isosteres can be obtained from measurements of a number of adsorption isotherms for various temperatures. Values of equilibrium pressure at constant coverage can be obtained from the isotherm and thus a graph of the isostere, $\log p$ against $1/T$, can be constructed from which the isosteric heat of adsorption can be determined by applying known thermodynamical relations. The heat of adsorption can also be measured directly with the aid of specially built calorimeters. When the mass of the metallic sample is sufficiently large, this is not a difficult problem. But for thin metallic films with comparatively small surface areas, a small total heat effect is obtained and this is difficult to determine accurately. In order to obtain a measurable temperature increase conveniently and accurately, specially constructed calorimeters with very small heat capacity, are used [46].

6.1.2.2 *Infrared spectrometry of adsorbed molecules*

Methods used for the investigation of adsorption equilibria and of thermodynamic parameters, while providing a good deal of knowledge of the adsorption process, often give only limited information on the nature of the bonding between adsorbate and adsorbent. The physical methods mentioned in Section 6.1.1 give information on the physical and chemical state of the adsorbent rather than adsorbate. Spectroscopic investigations of adsorbed layers particularly those on metallic surface — can, however, provide more information on the changes in molecules involved in the adsorbed state. Infrared absorption spectra and more recently Raman spectra have proved to be particularly useful in this regard. It is however necessary to point out that the vibration frequencies corresponding to the metal–admolecule bonding can be observed in only a few cases. In most investigations, only intra-molecular oscillations of admolecules, ad-radicals or ad-ions are accessible. However, it may be possible, by comparison with the corresponding bands of the free molecules, to identify and characterize adsorption complexes and in many cases to distinguish between physical and chemical adsorption. The occurrence in a given adsorption species of bands which are absent in a free molecule, indicates chemisorption.

Measurement of the spectra of adsorbed substances is complicated by several difficulties. Firstly, absorption of infrared radiation by a single adlayer is very small and in order to obtain a measurable adsorption it is necessary to arrange experimental conditions so that the infrared beam makes multiple passes through the adsorbed layer. Dispersed metals, as powders or deposited on supports, sometimes absorb the IR radiation more strongly than the molecules adsorbed on their surface. Fortunately, it has been found that the effect of the carrier and of the metal can be reduced to an acceptable level by using a very high degree of dispersion of metal and carrier, since in this situation and at the proper thickness of the absorbing sample, the adsorbate layers constitute a significant fraction of the optical path. Consequently, infrared transmission spectroscopy is well-suited to the investigations of adsorption on metals and highly dispersed solids.

Another experimental difficulty surrounds the need for high vacuum conditions for purifying and degassing specimens before adsorption and the need to correct for the IR absorption by the gaseous phase surrounding the sample. Effective pumping is difficult using absorption cells fitted with IR transmitting windows made of alkali halide crystals which cannot be adequately sealed to glass cells. The use of calcium fluoride and magnesium oxide as window materials have eliminated such problems.

Investigations of IR absorption spectra of molecules adsorbed on metals, conducted over almost thirty years, have provided numerous results which have been critically reviewed in several monographs [49, 50] and articles [51–53].

Adsorption of gases which constitute the most widely-used reactants in catalytic reactions on transition metals — carbon monoxide, hydrogen, the hydrocarbons and nitrogen, etc. — have been widely investigated by IR absorption methods. The carbon monoxide molecule in the gaseous state has two absorption maxima at wavelengths of 4.60 and 4.72 μm. In the adsorbed state on metals of groups IB and VIII, additional peaks are seen and analysis suggests that the additional absorption peaks correspond to the various adsorbed forms of CO — specifically the forms which differ according to the kind of bonding of the carbon atom with the metal

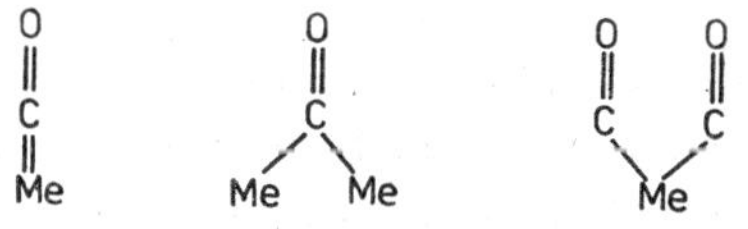

I II III

The linear form I, is believed to occur in the adsorption of carbon monoxide on the surface of iron, copper and platinum, the bridge form II — in the adsorption on nickel and platinum. In the case of adsorption of CO on rhodium all three forms I, II and III (two-centre form) have been observed. The non-metallic substrate (carrier) on which the metal is deposited also influences the ratio of each form.

The absorption spectra of hydrogen and deuterium adsorbed on platinum [54] contain two characteristic maxima at 4.74 μm and 4.86 μm, which have been assigned to Pt—H vibrations of linear (I) and bridge (II) forms

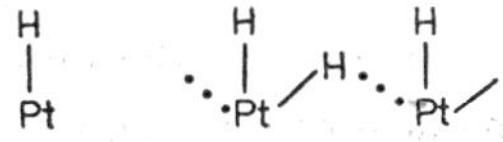

I II

However, other authors have subsequently claimed that the occurrence of these absorption bands depends on previous contact of platinum with oxygen [55]. It thus seems possible that hydrogen adsorbs in the infrared-active form only at places where surface oxide has previously formed. Spectra of hydrogen and deuterium adsorbed on rhodium have been obtained but not for hydrogen adsorbed on nickel.

There are also numerous investigations of the IR absorption spectra of hydrocarbons adsorbed on metals. As an example, work on ethylene on nickel with simultaneous adsorption of hydrogen can be mentioned. In this special case the analysis of absorption bands revealed the presence of stretching C—H vibrations (3.4–3.5 μm), similar to those in saturated hydrocarbons, and "scissors" motions of CH_2— (6.91 μm). This suggests the presence of the associative form of ethylene adsorption involving two nickel atoms

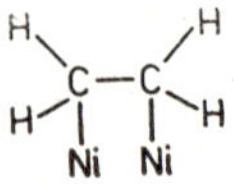

At higher temperatures (420 K) and in the absence of adsorbed hydrogen, a different absorption spectrum is obtained, which is ascribed to dissociative adsorption where one carbon atom can be bound to more than one nickel atom.

In the case of weaker, physical adsorption at low temperatures, the IR absorption spectra can give valuable information on the nature of the interaction of the adsorbate molecule with the adsorbent. Good examples of such investigations are the carefully-interpreted spectra of nitrogen adsorbed on highly dispersed nickel, palladium and platinum with crystallite diameters below 10 nm [56].

6.1.2.3 *The influence of adsorbed gases on the magnetic properties of highly dispersed metals*

The formation of a bond between the adsorbate and adsorbent involves electron transfer between the metal and adsorbate producing changes in the electronic states of the surface metal atoms. These changes can easily be detected and measured for ferromagnetic and paramagnetic metals. The specific magnetization of ferromagnetic and the magnetic susceptibility of paramagnetic metals depends directly on the spin states and the number of electrons in outer shells. Changes in the number of electrons with unpaired spins resulting from adsorption causes corresponding changes in these experimentally measured quantities.

Descriptions of the changes in magnetization of ferromagnetic metals accompanying the adsorption of hydrogen and other gases are presented in two monographs [57, 58], where references to detailed works can also be found.

In the adsorption of gases such as hydrogen, hydrocarbons, hydrogen sulphide, oxygen, etc. on nickel, a decrease of magnetization occurs, which is dependent on the degree of dispersion of the metal and on surface coverage

$$\frac{\Delta M}{M} = \frac{2N_a \varepsilon \beta}{I_{sp} V} + \left(\frac{N_s \varepsilon \beta}{I_{sp} V}\right)^2, \tag{6.27}$$

where $\Delta M/M$ is the relative decrease in magnetization, N_a is the number of molecules adsorbed on a single metal particle, ε is the number of electrons transferred to metal in the adsorption of one molecule or one atom, β is the Bohr magneton, I_{sp} is the spontaneous magnetization per unit volume and V is the volume of a nickel particle. The quadratic term is usually small compared with the linear term and in most cases can be neglected. Approximately linear dependence of $\Delta M/M$ on N_a indicates that the nature of adatom binding does not undergo substantial changes with increasing coverage. The decrease of magnetization also shows that a decrease in the unpopulated d-states in nickel occurs, hence the decrease of the magnetic moment. It may be concluded therefore that only covalent or ionic $Ni—H^+$ forms of adsorption bonding of hydrogen are possible and that the ionic $Ni—H^-$ form does not occur.

Magnetization changes following adsorption have been used to study ferromagnetic metals and their alloys and have given remarkable results for nickel and its alloys. Knowing the values N_a and V in eq. (6.27) it has been possible to determine the number of electrons per admolecule involved in the chemisorption bond and thus to attempt to establish the bonding model. This method has been used also in more recent investigations of adsorption of hydrocarbons and hydrogen sulphide on nickel catalysts [59]. Dissociative adsorption and complete cracking of these hydrocarbons at higher temperature (320–470 K), has been established.

Magnetic investigations of the adsorption of hydrogen and hydrocarbons on nickel have recently been repeated by van Meerten *et al.* [60] and the conclusions drawn from these measurements were essentially identical to those mentioned above with only minor corrections. In further investigations of this kind, the adsorption of ethylene on silica-supported Cu-Ni alloy has been studied [61] with the result that, depending on the composition of the adsorbent and its temperature, ethylene adsorbs in two, so-called π and di-σ forms

π-form (77K) di-σ form (243-293K)

6.1.2.4 Adsorption-induced changes in the electrical conductivity of dispersed metals

Adsorption-induced changes of the electronic structure of the surface atoms will affect the electric conductivity of highly dispersed metals. Changes in the electrical conductivity are most conveniently measured for thin metallic films evaporated in high vacuum and then exposed to the gas. In most cases a decrease of conductivity is observed. The effects are not very large — at a film thickness of the order of 10 nm they do not exceed 1–2% — but are easily measurable. Interpretation of these effects is not a simple matter, because there are many possible mechanisms which lead to an increase of electrical resistivity due to the presence of admolecules on the surface of a metal. The electrical resistance of thin metallic films can be expressed by the formula

$$R = \frac{Af(d/l_0)}{N^{3/2}dl_0}, \tag{6.28}$$

where A is a constant, d is the film thickness, l_0 is the electron mean free path and N is the number of conduction electrons per unit volume. This formula shows that changes in the conductivity due to adsorption can arise from changes in the number of conduction electrons, in the effective film thickness, or from changes of the electron mean free path. Simultaneous changes of more than one of these quantities is also possible. The number of conduction electrons can decrease as a result of the involvement of the electrons of the surface atoms in bonds to the adsorbate. In this way the layer of surface atoms can undergo "demetallization", i.e. the effective film thickness decreases. Similarly, the presence of an adsorbate layer can change the mobility of the conduction electrons (which is directly related to l_0). Separation of these three factors is, in general, impossible and therefore the mechanism of the conductivity decrease due to adsorption has not been finally clarified.

For the adsorption of hydrogen and CO on many metals, the dependence of the relative resistance change $\Delta R/R$ on the coverage is linear at small coverages. At coverages above half a monolayer, however, there is a tendency for saturation and sometimes $\Delta R/R$ even decreases again. These effects can arise from the mutual interaction of admolecules and from different types of adsorption. The experimentally established facts do not always accord with results obtained by other methods — for instance IR absorption, desorption spectra and surface potential measurements — and in many cases cannot be unambiguously interpreted.

It may be necessary to point out that the occurrence of a decrease in the electric conductivity due to adsorption does not necessarily indicate the chemical character of the bonding. Similar, though quantitatively smaller, effects are also observed in cases where physical adsorption undoubtedly occurs — for example, adsorption of the inert gases [62].

6.1.2.5 Other methods

The methods described in Sections 6.1.1.2 (Auger electron spectrometry), 6.1.1.3 (photoelectron spectroscopy), 6.1.1.5 (thermal desorption) and 6.1.1.6 (the diode and vibrating capacitor methods), can be also applied to investigations of adsorption on highly dispersed metals; in particular when the metal is in the form of a thin film deposited on a non-metallic substrate.

Highly dispersed metallic adsorbents however, create experimental difficulties. For adsorption measurements to be physically significant, the metal surface and the adsorbate should be as clean as possible, and therefore some cleaning procedures are necessary. It is generally not possible to use conditions and procedures which would be suitable for single crystal surfaces such as heating in ultra-high vacuum, ion bombardment, annealing, etc., because these would alter the degree of dispersion of the metal.

Furthermore, concentration of structural defects on the surface of a highly dispersed metal is usually much larger than on single crystal surfaces. Consequently the results obtained for highly dispersed metals are rarely valid as a basis for general conclusions but in view of the great practical significance of adsorption on metallic powders, thin films and supported metallic catalysts, such investigations are useful and should not be underestimated.

6.1.2.6 Adsorption on the smallest metallic particles

Adsorption measurements on highly dispersed metals deposited on non-metallic supports have shown that there are significant quantitative differences in the adsorption of a given adsorbate on massive and highly dispersed metals. The number of admolecules per metal atom of the surface of a macroscopic crystal, does not exceed 1, as a rule. It has been established by surface spectroscopy (AES) and desorption measurements that the maximum coverage of a (111) platinum surface by chemisorbed CO is 0.68 [63], and for (100) platinum, 0.75 CO molecules per surface platinum atom [64]. It has been observed however, that the adsorption capacity of highly dispersed nickel for methane is higher than that of large crystallites [65].

Similarly, the ratio of adsorbed CO molecules to rhodium atoms in flat clusters ("rafts") of the latter deposited on Al_2O_3 is equal to 2 for the smallest clusters (containing up to 6 atoms) and for clusters containing up to 36 atoms the ratio drops to 1.5–1.6 [66].

Such high ratios of the number of adsorbate molecules to surface atoms seem to be connected with geometrical differences between the extended surface planes of macroscopic crystals and those which constitute the surface of the smallest crystallites. In the latter there are also more atoms on edges and corners, which can be bound to more than one adsorbate molecule. The influence of the modified electronic structure of small crystallites and clusters compared with larger crystals, and the further modification induced by the support can also play a role in this enhancement of the adsorption capacity.

6.1.3 Application of the methods of quantum chemistry

The structure of adsorption complexes — namely entities composed of adsorbed molecule and a few atoms of the adsorbent — has been investigated theoretically with the aid of quantum chemical methods (see Chapter 4). Many of the early calculations on chemisorption were carried out by the EH method, while the more advanced quantum chemical methods came into use more recently. In what follows, the account of the results of these methods applied to the most common chemisorbed systems follow the same historical progression.

6.1.3.1 The Extended Hückel method

Robertson and Wilmsen [67] considered adsorption of CO molecules on the (100) face of nickel by applying the EHM to an adsorption complex centred on eight Ni atoms. The results of their calculations showed that the energetically favoured forms of the chemisorbed CO molecules are the "on top" (I) and bridged (II) forms with corresponding adsorption energies of 2.39 and 2.52 eV, respectively

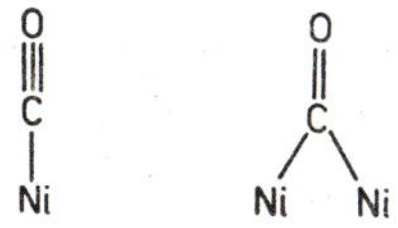

I II

In both these configurations the bonding of the molecule is through a carbon atom. The small difference of binding energy between these two forms suggests that they can be simultaneously present on the Ni (100) surface. Moreover, according to the authors' calculations, the double adsorption sites for form (II) are separated by a rather small (0.2 eV) potential barrier, so that considerable mobility of CO admolecules in this form is possible. The authors have also established that a random arrangement of molecules adsorbed as in (II) is the most probable, and that the admolecules are not in contact with each other. This arrangement corresponds to a saturation coverage of one CO molecule for two nickel atoms, in agreement with experimental data. The calculated values of the binding energies are also fairly close to those earlier observed.

Among the earlier attempts to apply the EH method to adsorption on metals were the calculations of the adsorption energy of the hydrogen atom on the 13-atoms cuboctahedral Ni-cluster, as well as smaller clusters, obtained by truncation of the latter, thus simulating configurations corresponding to (100), (110) and (111) Ni-planes [68]. The most important result of this work is, that adsorption of the hydrogen atom on three- and four-fold adsorption sites (i.e. at tetrahedral and octahedral holes) is energetically less favourable than adsorption on top of a single atom, for all three crystallographic planes. It has also been found that the adsorption energy decreases a little when an increased number of Ni atoms around the Ni—H "surface molecule" is included in the calculation. This suggests a repulsive interaction between the surrounding Ni atoms. The results also indicate a significant contribution from the 4*s* orbitals of nickel in the formation of the bond with hydrogen, which is consistent with the fact that the bond strength between hydrogen aud copper — with filled *d* orbitals — in only a little weaker than the bond to nickel.

Adsorption of hydrogen on the (100) face of tungsten (bcc) has also been treated by the EH method by Anders *et al.* [69]. Nine- and twelve-atomic clusters of tungsten atoms representing the (100) plane and the "on top" two-fold and five-fold adsorption sites have been considered. The highest binding energy has been obtained for the single "on top" configuration namely, 3.0 eV at the minimum W-H distance of 0.165 nm. The binding energy at the other two sites was smaller: 2.7 eV for a two-fold (bridge) site and 1.8 eV for a five-fold site. It is clear that the difference in binding energies between single and two-fold sites is not large so that the diffusive movement of a hydrogen adatom from one single site to another, through a two-fold site becomes possible. The experimentally-determined activation energy of diffusion (0.26 eV), [70] is in good agreement with this calculated difference.

Calculations for the adsorption of nitrogen atoms on the same sites, were carried out in a similar way [71] and showed that adsorption is most probable on a five-fold site with a binding energy of 3.25 eV, compared with 3.15 eV for a two-fold and 2.85 eV for a single site.

The most complete extended Hückel study of adsorption was carried out by Anderson and Hoffmann [72] who studied the interaction of several diatomic molecules — Li_2, B_2, C_2, N_2, O_2 and F_2 as well as CO and NO with nine-atom clusters representing the (100) surface planes of nickel and tungsten (see Fig. 6.12).

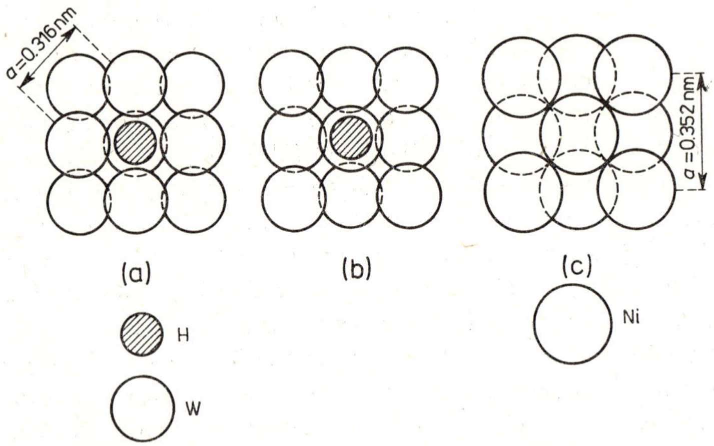

Fig. 6.12 — Nine-atom clusters representing (100) plane of tungsten (a), (b) and nickel (c). The adsorbed hydrogen atom in position (a) has five nearest neighbours, and one nearest neighbour in (b).

The results of these calculation suggested that:

(i) there is a strong tendency for charge transfer between the metal and adsorbate, proportional to the difference of their electronegativities,

(ii) the adsorbed molecules tend to dissociate as a result of strong Coulombic attraction with filling of antibonding orbitals,

(iii) the interactions are strong, regardless of the position and orientation of the adsorbate, i.e. the adsorbing surface is to a large extent homogeneous,

(iv) as a result of accumulation of charge at "corners" and "steps", there is an increased tendency for adsorption and dissociation of molecules at these sites,

(v) the photoemission spectra (UPS) for CO on W (100) and Ni (100) obtained experimentally, are in good agreement with the density of states curves resulting from calculations of the electronic structure.

A recent extended Hückel study of chemisorption and catalysis on metallic clusters, carried out by Halachev and Ruckenstein [73] was aimed at an explanation of the effect of additions of copper, sulphur, silver and chlorine on the catalytic activity of platinum and silver clusters. Model clusters were composed of either ten atoms (7 atoms in the (111) plane with three below), or nine atoms (five atoms in a (100) plane and four in the plane below) (see Fig. 6.13). It was been found that additions of

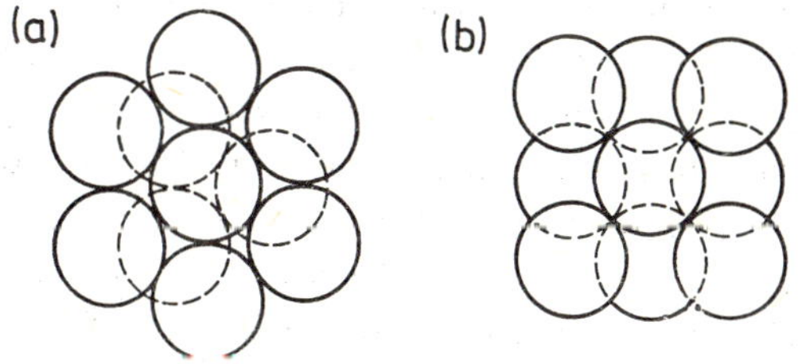

Fig. 6.13 — (a) Ten-atom cluster representing the (111) plane of a fcc metal, (b) nine-atom cluster representing the (100) plane of a fcc metal.

metallic Ag and Cu modify the population of Pt—Pt bonding electrons only slightly, while the number of electrons belonging to individual atoms increases strongly. Thus the electron density on the cluster surface increases due to transfer of electrons from Cu and Ag, thus accentuating the transfer of electrons from the cluster into antibonding π^*-orbitals of the adsorbed ethylene. A further consequence is an enhancement of chemisorption of oxygen, and an increase in the rate of oxidation of ethylene and of carbon monoxide. Sulphur present in the cluster causes inhibition of these reactions since it acts as a strong electron acceptor and competes for electrons, thus destabilizing the molecular orbitals and decreasing the oxygen chemisorption. Inhibition by sulphur is stronger than the promoting effect of silver and copper.

6.1.3.2 Other quantum chemical methods

Other, more rigorous quantum chemical methods have been applied more recently to problems of chemisorption. The CNDO method has been used by Blyholder to calculate the interaction of hydrogen atoms with 10-atom clusters, representing (100), (110) and (111) nickel surfaces [74]. The results of these calculations showed that the hydrogen atoms are more strongly bonded at multiple-bonding sites than at single "on top" sites and that the binding energy on the (111) face is highest. It has been shown that the *d*-orbitals of nickel play only a minor role in the bonding compared with *s* and *p* orbitals. The adsorbed hydrogen atom has an excess negative charge and the calculated bond length was found to be 0.16 nm compared with the experimental value of 0.147 nm. The most probable location of the hydrogen adatom predicted by this work is at variance with that suggested by the EH calculations of Fassaert *et al.* cited above [68]. However, the results of Blyholder seem more credible, since the CNDO method gives the equilibrium positions of atoms, while the EH method is much less sensitive to atom positions.

The SCFXα-SW method has also been used in calculations of the energy levels of chemisorption complexes. One of first calculations to be carried out using this method was for the Ni—O system [75], where the similarity of the electron level scheme obtained with the ultraviolet photoelectron spectrum (UPS) obtained for a nickel surface with a high concentration of adsorbed oxygen, has been established. Calculations have been performed for the octahedral complex NiO_6^{10-} rather than for the Ni_xO complex, which is justifiable since chemisorption of oxygen on nickel at moderate temperatures leads to oxidation and formation of a surface layer of nickel oxide, NiO.

SCFXα-SW calculations for a Ni_5O complex with four-fold adsorption sites and comparison of the results with UPS spectra obtained at low oxygen coverage [76] have been reported by Batra and Robaux [77]. Assuming the Ni—Ni distance to be the same as in the massive metal, they calculated the electron energy levels for the cluster, and found that the levels calculated for the chemisorbed oxygen were in good agreement with a broad maximum in the UPS spectrum which appeared after exposure of the nickel to oxygen. It was also found that the interaction with the fifth nickel atom (in the second atom layer) is negligible, and that the chemisorption bond is formed with contributions of *p* oxygen orbitals and *d*, *s* and *p* orbitals of nickel. The Ni_5CO chemisorption complex has also been investigated by SCF-Xα-SW techniques. Calculations showed that it is sufficient to consider five nickel atoms: an increase of the number of nickel atoms in the complex to nine did not significantly alter the results [78].

In this latter study of Ni_5CO, the adsorption bonding was described on the basis of σ orbitals of CO and *s*, *p* and *d* orbitals of nickel. This is not completely in agreement with Blyholder's calculations obtained for a similar system with the CNDO method — Blyholder found that the *d* orbitals of nickel participate only slightly in the bonding [79]. The latter calculations refer however, to a $Ni_{10}CO$ complex, representing adsorption on the (111) face of nickel.

Adsorption of hydrocarbons on nickel has also been treated theoretically. Calculations of the electronic structure of acetylene, ethylene, propylene and benzene adsorbed on nickel have been carried out by an *ab initio* SCF LCAO (Self-Consistent Field Linear Combination of Atomic Orbitals) method and the results compared with the photoelectron spectra of the free and adsorbed molecules [80]. From this comparison it was concluded that adsorption of these hydrocarbons is via bonds involving the *d* orbitals of nickel and π orbitals of the hydrocarbon molecules, as in metal-organic π-complexes. There is no rehybridization (for instance $sp^2 \rightarrow sp^3$ in ethylene) and no σ-bond formation.

The more recent theoretical investigations of adsorption on metals have tended to apply more rigorous SCF and *ab initio* quantum chemical methods, especially for cases of adsorption of hydrogen [81] and carbon monoxide [82, 83]. However, since these methods involve difficult and very long calculations, they are usually performed for adsorption complexes consisting of only a small number of metal atoms. Only qualitative agreement of the results with experimentally obtained UPS spectra for highly dispersed adsorbents can be expected in most cases.

6.2 CATALYSIS ON HIGHLY DISPERSED METALS

6.2.1 Catalysis on solid surfaces

It is well known that the rate of a chemical reaction may be accelerated by the introduction of an additional reagent — the catalyst — to the reacting system. This produces *active centres* which participate in the *active complex* (transition state) of the slowest elementary reaction step. The formation of the active complex with the participation of active centres of catalyst changes the reaction path in such a way that the activation energy of the slowest step is lowered and the entire reaction proceeds with an increased rate. Since the active complex decomposes very quickly in the course of the reaction, the quantity and chemical composition of the catalyst remain unchanged after the completion of catalyzed reaction. Thus the catalyst does not appear in the material balance of the chemical process and very small quantities can produce the desired chemical change under appropriate conditions of pressure, concentration of reactants and temperature.

Catalytic processes are, of course, of great importance in the chemical industry, and many of the very large scale industrial chemical processes are almost totally catalytic.

A number of chemical elements and compounds can act as catalysts. It is an old custom to divide catalytic processes and catalysts into cases, where the catalyst and other reactants constitute a single homogeneous phase — homogeneous catalysis — and those where the catalyst forms a separate (usually solid) phase — heterogeneous catalysis. In the latter case the active centres are localized at the reagent–catalyst interface. The extent of this interface and the concentration of active centres thus become crucial parameters governing the efficiency of the catalyst.

Efficiency is characterized by two quantities: the *activity* expressing the rate of a catalyzed reaction compared to a non-catalyzed one under the same conditions, and *selectivity*, i.e. the ratio of the rate of the desired reaction to the rates of all others reactions proceeding in the given reactant system. The rate of a catalytic reaction is precisely and conveniently expressed in terms of the *turnover number*, i.e. the number of reactant molecules reacting per catalytic centre in unit time. For heterogeneous catalytic reactions the turnover number varies over wide limits (for instance, 10^{-3} to 10^2 molecules s^{-1} per centre). In order to increase the number of active centres participating in a heterogeneous catalytic reaction, catalysts with a large specific surface area and a high degree of dispersion are used. For example, the specific area, in many cases equals a few hundred $m^2 \cdot g^{-1}$, and the degree of dispersion may be near unity. Methods of manufacturing catalytically active metals in such highly dispersed forms have already been described in Chapter 2.

Heterogeneously-catalyzed reactions proceed on the surface of solids of various kinds: metals, semiconductors and insulators. In the majority of cases they are inorganic substances. Since there is a vast number of practically and scientifically important catalytical reactions, they cannot be characterized here even in most general way. The interested reader can find good introductions in appropriate introductory books and monographs [84–87].

The common feature of all heterogeneous reactions is that chemisorption of at least one reactant at the catalyst surface is a necessary condition. Thus some information on the chemisorption of gaseous reagents on the dispersed metals and transformations of chemisorbed molecules via active complex formation is given in following paragraphs.

6.2.2 Catalytic activity of metals

Metals account for a large proportion of the catalysts used in chemical technology. It has been recognized for many years that an indispensable condition for a metal to act a catalyst is that reactants chemisorbed on the surface should not be too strongly bonded (Sabatier's rule).This means that the chemical affinity between the active centres of the catalyst and the reactants should not be too great. The nature of active centres is however not always easy to determine, and thus theoretical predictions and calculations of this affinity are rarely possible. Nevertheless, there have been numerous attempts to search for a relation between the experimentally accessible general properties of a catalyst, and its activity. This has led to notions of "geometric" factors — involving the structure of the surface — and "electronic" factors (electronic structure) being responsible for the catalytic properties of solids. The role of these factors has been well-characterized for pure metallic catalysts, since in these cases our knowledge of the geometric surface structure and electronic properties are most complete.

6.2.2.1 Geometric factors in catalysis on metal surfaces

If the activated complex (transition state) in an elementary reaction is formed with the participation of more than one surface atom, a special configuration may be required and, in the absence of such a configuration, the active complex is not formed and thus catalytic effects do not occur. When only one surface atom is engaged in the activated complex, the distance and configuration of neighbouring atoms are, clearly, unimportant.

The former situation may lead to special requirements for certain catalytic reactions such as the number, distance and symmetry of the surface atoms. Such special configurations were postulated some time ago by Balandin [88] (multiplets) and Kobozev [89] (ensemble). More recently, catalytic reactions requiring such special arrangements of surface atoms have been called *demanding* or structurally-sensitive reactions. The reactions which do not require the presence of such special arrangements of atoms are then *undemanding*, structurally-insensitive or facile reactions [90]. In Table 6.1 some examples are given of structurally-sensitive and insessitive catalytic reactions, proceeding on the surfaces of transition metals.

Table 6.1 "DEMANDING" AND "FACILE" REACTIONS

Reaction	Catalyst	Other catalysts	Remarks	Ref.
Demanding reactions:				
Decomposition of ammonia: $2NH_3 \rightarrow N_2 + 3H_2$	W(111)	W(100) and (110)	The reaction rate on W(111) about 10 times greater	[91]
Ethane hydrogenolysis	Rh	—		[92]
Hydrogenolysis of *n*-hexane	Rh, Pt		Dependent on particle size	[93]
Ethane hydrogenolysis	Ni-Cu alloy (up to 10% Cu)	Ni	Turnover number for Cu-Ni alloy 10^3 times smaller than for pure Ni	[94]
Cyclopentane hydrogenolysis	Rh/Al_2O_3		Reaction rate dependent on dispersion	[95]
Facile reactions: $CO + \frac{1}{2} O_2 \rightarrow CO_2$	Various crystallographic planes of Pd and Pt	Various transition metals and their oxides	No differences in the chemisorption and catalytic activity between polycrystalline Pd and its (111), (110) and (100) planes.	[96]
Cyclopropane hydrogenation	Pt/Al_2O_3 Pt/SiO_2			[97]
Hydrogenation of cyclohexane benzene	Pt/support			[98]
H_2-D_2 exchange, cyclohexane dehydrogenation	Cu-Ni alloy (up to 90% Cu)			[94]

Some of the metallic catalysts cited in the Table 6.1 were prepared such that samples with various values of the degree of dispersion have been obtained, which could give rise to variable numbers and types of active centres. The presence and concentration of various active centres and their dependence on the metal dispersion has been shown by van Hardeveld and Hartog [99]. Thus a lack of dependence of the reaction rate on the degree of dispersion indicates that the reaction is structurally insensitive. The inverse statement may not always be true, because the dependence of the catalytic activity on the dispersion of the catalyst does not exclude a simultaneous dependence on factors connected with the change of electronic structure with dispersion. Effects due to the interaction of the metal with its support, which is dependent on the degree of dispersion may also be important.

6.2.2.2 *Electronic factors*

Although the idea that the catalytic activity is dependent on the electronic structure of a catalyst has been current for more than four decades, nonetheless the concept is still rather qualitative in nature. An attempt to quantify this notion by Pauling [100] lead to the introduction of the so-called "*d*-character" of transition metals and the recognition of its relation to catalytic activity in reactions involving hydrogen and hydrocarbons. The *d*-character is defined as the proportion of *d* orbitals involved in the formation of hybridized *dsp* orbitals. The higher the *d*-character, the smaller is the fraction of free *d* orbitals. Numerically it varies from almost 39% for chromium to 50% for rhodium. An increase in *d*-character is accompanied by an increase in the specific activity of a metal surface in the hydrogenation of hydrocarbons [101]. The *d*-character can be calculated from the relation

$$R_1 = 1.825 - 0.043z(1.60 - 0.1z)\delta\ , \qquad (6.29)$$

where R is the radius of the single-bonded metal atom, z is the number of electrons in excess of closed shell of the preceding noble gas atom, and δ is the *d*-character. The relation between the metallic radius and the *d*-character suggests that the electronic factor is related to the geometric factor. On the other hand, it is easy to see that a correlation between the *d*-character and the catalytic activity does not explain differences in activity between the various surface planes of the same metal (see Table 6.1), i.e. it does not account for the difference between the demanding and facile reactions.

The electronic factor was later redefined by Bond [102] in terms of surface metallic orbitals preserving the directionality and other properties of the orbitals in bulk metals. The absence of next neighbours (ligands) results in coordinative unsaturation of the surface atoms, i.e. the occurrence of unsaturated bonds. The quality and quantity of these vacant orbitals, which can be filled by electrons from the adsorbate molecules represents the "electronic factor" in this approach and the relation with the geometric factor is easier to recognize. Both of these factors can simultaneously govern the formation of unstable activated complexes, the further transformation of which leads to the final products of catalytic reactions.

6.2.2.3 Metal alloys as catalysts

Metallic alloys (Chapter 2) have several advantages over the monometallic catalysts such as improved thermal stability, enhanced activity and selectivity and, in some cases, easier regeneration (without loss of activity due to sintering) after their use in reactions involving hydrocarbons [103]. Probably the oldest commonly-used binary alloy catalyst is platinum–rhodium in the form of wire gauze used in the oxidation of ammonia in the manufacture of nitric acid. Another example of alloy catalysts with a long history are the Raney metals (Chapter 2) which always contain a certain amount of metallic aluminium. (So-called "pure metals" used as catalysts often contain small amounts of other metals and can be considered as diluted alloys).

Early interest in alloy catalysts was mainly concentrated on alloys of the group VIII metals with elements of group IB. These alloys are mostly solid solutions (at least over certain composition ranges) and changes in the electron concentration in the *d*-bands of these alloys (as compared to pure metals of group VIII and the effect on catalytic activity in reactions involving hydrogen and hydrocarbons have been investigated in the hope of creating some support for the concept of an electronic factor.

These investigations have however, produced results which contradict the notion that the bulk properties of transition metals govern their chemisorptive and catalytic properties. Even minute additions of IB and IIB metals like Cu [104, 105] and Zn [106, 107], which modify the filling of the *d*-band only slightly, have a dramatic effect on the catalytic activity of the group VIII metals in hydrogenation reactions.

An explanation for this fact became possible, when it was recognized that the chemical composition of the *surface* of the alloy can differ significantly from that of bulk phase. This is believed to be mainly the result of two effects:

1. On thermodynamic grounds — specifically Gibbs law — it follows that, in the equilibrium state, an excess of the component which lowers the free surface energy (surface tension) will accumulate at the surface. The surface region thus becomes relatively rich in the component having a lower sublimation energy and/or a larger atomic radius. In the case of the alloys mentioned above Cu and Zn are segregated to the surface and their activity in the chemisorption of hydrogen and hydrocarbons is near zero.

2. In a number of cases, it has also been observed that the surface becomes enriched with the component which has the greater chemical affinity for the gases which are in contact with the surface (oxygen, hydrogen, carbon monoxide). For instance, in the Ag-Pd alloys the surface is enriched in Ag in vacuum, in accordance with Gibbs law; however, when these alloys are exposed to carbon monoxide, an excess of Pd [108] appears on the surface. The Au-Pd alloys showed Pd-enrichment of the surface when exposed to oxygen, since the affinity of palladium for oxygen is much greater than that of gold [109]. Similar results were obtained for a Ni-Au alloy, the surface of which is enriched in gold in vacuum, but after exposure to oxygen the surface concentration of gold decreases appreciably [110].

Enrichment of alloy surfaces by one of the components predicted by thermodynamic arguments [111, 112] can be verified experimentally by Auger electron spectroscopy [109, 110] and photoelectron spectroscopy [113]. Measurement of the specific chemisorption of gases such as hydrogen, oxygen and carbon monoxide and of the surface potential can also be used here.

There is an established tendency to distinguish between alloy catalysts with larger particles, which can be used either unsupported or supported on a non-metallic carrier, and very highly dispersed, supported catalytic alloys — bimetallic or multimetallic clusters [103]. Surface enrichment by one of the components seems to occur even in very small three-dimensional clusters [114]. There are also other effects, such as an increase in miscibility, mentioned in Chapter 2, of metals which do not form solid solutions in the bulk — for example, the Cu–Ru and Cu–Os systems. This effect has been established on the basis of chemisorptive properties and catalytic activity, since diffraction and surface spectroscopy methods are less suitable for metals in a state of extreme dispersion.

The attractive catalytic properties of bimetallic clusters have been described and discussed in some depth in a number of patents and publications (see for instance [103, 115, 116]. The beneficial changes in activity and selectivity compared with the elemental metals best reveal themselves in demanding catalytic reactions involving multicentre adsorption of hydrocarbons, or other organic compounds, and this accounts for their increasing application as industrial catalysts.

6.2.2.4 Metallic molecular clusters in catalysis

Over the last 15 years a large number of investigations have been carried out on the structure and catalytic properties of a group of chemical compounds similar to metallic clusters and catalysing the same chemical reactions. They are composed of a small number of metal atoms bonded together and having similar geometrical configurations to those of the metal clusters, but with other groups of atoms or molecules, located on the periphery of the metallic "core". They thus resemble metallic clusters with the non-metallic atoms or molecules "adsorbed" onto the surface. Examples of such compounds — *metallic molecular clusters* — are listed in Table 6.2. According to traditional chemical nomenclature these

Table 6.2 METALLIC MOLECULAR CLUSTERS

Number of metal atoms	Formula	Geometric arrangement of metal atoms
1	2	3
3	$Fe_3(CO)_{12}$	
	$Ru_3(CO)_{12}$	
	$Os_3(CO)_{12}$	Triangle
	$Pd_3(CNR)_6$	
4	$CO_4(CO)_{12}$	
	$Rh_4(CO)_{12}$	
	$[Fe_4(CO)_{13}]^{-2}$	Tetrahedron
	$Ni_4[CNC(CH_3)_3]_7$	
	$[Re_4(CO)_{16}]^{-2}$, $[Fe_4(CO)_{13}H]^-$	"Butterfly" *
	$Pt_4(O_2CCH_3)_8$	Square

1	2	3
5	$Os_5(CO)_{16}$ $[Ni_5(CO)_{12}]^{-2}$ $[Fe_5\underline{C}(CO)_{15}]^{-2}$	Trigonal bipyramid Trigonal bipyramid Square pyramid **
6	$Rh_6(CO)_{16}$, $[CO_6(CO)_{14}]^{-4}$ $[Rh_6(CO)_{15}\underline{C}]^{-2}$, $[Pt_6(CO)_{12}]^{-2}$ $Os_6(CO)_{18}H_2$	Octahedron Trigonal prism Capped square pyramid ***
7	$[Rh_7(CO)_{16}]^{-3}$, $Os_7(CO)_{21}$	Capped octahedron
8	$[CO_8(CO)_{18}\underline{C}]^{-2}$ $Ni_8(PC_6H_5)_6(CO)_8$ $Rh_8\underline{C}(CO)_{19}$	Antiprism Cube Bicapped trigonal prism
9	$[Ni_9(CO)_{18}]^{-2}$, $[Pt_9(CO)_{18}]^{-2}$	Double trigonal prism
12	$[Pt_{12}(CO)_{24}]^{-2}$	Triple trigonal prism
13	$[Rh_{12}\underline{Rh}(CO)_{24}H_e]^{-2}$	Truncated hexagonal bipyramid with one Rh atom inside
14	$[\underline{Rh_{13}Rh}(CO)_{25}]^{-4}$	Five-fold capped bcc cube
15	$[(H_2O)Rh_{14}\underline{Rh}C_2(CO)_{28}]^{-}$ $[Pt_{15}(CO)_{30}]^{-2}$	Four-fold capped pentagonal prism Four-fold twisted trigonal prism
18	$[Pt_{18}(CO)_{36}]^{-2}$	Five-fold twisted trigonal prism
19	$[Pt_{17}\underline{Pt_2}(CO)_{22}]^{-4}$	Bicapped double pentagonal prism
30	$[Pt_{30}(CO)_{60}]^{-2}$	Nine-fold twisted trigonal prism

* The "butterfly" configuration formed by stretching one side of the tetrahedron.

** C-atom inside the polyhedron.

*** "Capped" means one additional atom on the face of polyhedron.

compounds may also be termed — *polynuclear metal complexes* — and the groups on non-metallic atoms bonded to the metal — *ligands.*

As can be seen from the Table 6.2, most of the molecular metal clusters currently known contain CO molecules as ligands; but hydrogen (hydride ion) and organic radicals are also common. It was shown by nuclear magnetic resonance measurements that these ligands are not strongly bound to the metal core, and are mobile to an extent comparable with the mobility of chemisorbed molecules.

Another similarity of molecular metal clusters to pure metallic clusters is that the geometric arrangements of the core atoms, is in many cases, similar to polyhedra found in fcc, hcp and bcc lattices. Thus tetrahedra, octahedra, cubes, hexagonal bipyramids and polyhedra with five-fold symmetry axes — pentagonal pyramids, prisms and icosahedra — are observed. However, other geometric forms also occur due to deformation (relaxation) of the metal–metal bonds caused by the presence of ligands.

The geometrical and electronic structures of metal molecular clusters has been reviewed extensively [117, 118, 119, 120, 121]. In some of these reviews their similarity to "bare" metal clusters and metallic surfaces are emphasized as well as their analogous catalytic activity.

Many investigations carried out in last years or so have shown that metal molecular clusters can act as catalysts, activating the carbon monoxide, hydrogen, oxygen and hydrocarbon molecules in homogeneous media and at temperatures much lower than those at which supported metal catalysts are used. The $Ni_4[CNC(CH_3)_3]_7$ cluster, for instance, acts as a catalyst in the selective hydrogenation of acetylene-type hydrocarbons to give *cis*-butenes, in trimerization reactions of acetylene to give benzene, and in the dimerization of butadiene to cyclooctadiene at room temperature [122]. Carbonyl clusters $Ir(CO)_{12}$, $Rh_4(CO)_{12}$ and $Rh_6(CO)_{16}$ catalyse the synthesis of hydrocarbons from carbon monoxide and hydrogen (Fischer–Tropsch reaction) at 433 K in molten salts [123]. Even more numerous are examples of the catalytic activity of carbonyl clusters deposited on high-surface area oxides like Al_2O_3, and MgO in the Fischer–Tropsch synthesis of hydrocarbons and the water gas shift reaction $CO + H_2O \rightarrow H_2 + CO_2$; both of which are important in industrial chemistry [124]. The water gas shift reaction is also catalysed by clusters of $H_4Ru_4(CO)_{12}$ and $H_2Ru_4(CO)_{13}$ at 373 K and 0.1 MPa CO pressure [125].

Carbonyl clusters of rhodium [126, 127, 128], ruthenium [129] and osmium [130, 131] supported on high surface area Al_2O_3 have been investigated by means of infrared absorption and Raman spectroscopy, gas analysis, isotopic exchange and electron microscopy with the objective of establishing their composition and structure over the temperature range at which they exhibit catalytic activity in reactions involving carbon monoxide. These investigations have shown that although carbonyl clusters undergo decomposition (decarboxylation) at higher temperatures, giving active species with different CO/metal ratio, the decomposition is reversible and the original composition can be restored at lower temperatures and sufficiently high CO pressure. Even microcrystallites of ruthenium, obtained by complete decomposition, can be converted to the original complexes in this way [128].

Molecular metal clusters used as catalysts can also be deposited ("anchored") on the surface of organic polymers and silica gels by chemical bonds [132, 133]. Iridium and osmium molecular clusters prepared thus, contain phosphine ligands as well as CO and catalyse the hydrogenation of ethylene and acetylene and isomerization of hydrocarbons.

Iron carbonyls deposited on high area Al_2O_3 and MgO by impregnation from solution form carbonyl-hydride cluster $[HFe_3(CO)_{11}]^-$ which selectively catalyses the formation of low molecular weight olefins in the Fischer–Tropsch synthesis [134].

There are also other examples of the preparation of highly dispersed, supported transition metal catalysts by the decomposition of molecular clusters containing these metals in vacuum or in an atmosphere of hydrogen at temperatures around 473 K. These catalysts were found to be active in the synthesis of methanol, methane and other hydrocarbons [135] from H_2 and CO at low temperature and pressure.

Investigations of molecular cluster catalysts, which have developed rapidly in recent years, offer new possibilities of a deep insight into the mechanism of reactions catalysed by metals and new information leading to a unified approach to the problems of homogeneous and heterogeneous catalysis.

REFERENCES

[1] R. Gomer, *Chemisorption*, in *Surface Science*, Lectures presented at International Course, Trieste 16 Jan.-10 April 1974, International Atomic Energy Agency, Vienna 1975, Vol. II, p. 63.
[2] C. J. Davisson and L. H. Germer, *Phys. Rev.*, **30**, 705 (1927).
[3] E. J. Scheibner, L. H. Germer and C. D. Hartman, *Rev. Sci. Instr.*, **31**, 675 (1960); L. H. Germer and C. D. Hartman, *ibid.*, **31**, 784 (1960).
[4] J. J. Lander, J. Morrison and F. Unterwald, *Rev. Sci. Instr.*, **33**, 782 (1962).
[5] R. L. Park and H. E. Farnsworth, *Rev. Sci. Instr.*, **35**, 1592 (1964).
[6] *LEED — Surface Structure of Solids*, Ed. M. Láznička, Union of Czechoslovak Mathematicians and Physicists, Praha 1972, Vols. 1 and 2.
[7] J. B. Pendry, *Low Energy Electron Diffraction*, Academic Press, London–New York 1974.

[8] G. Ertl and J. Küppers, *Low Energy Electrons and Surface Chemistry Monographs in Modern Chemistry*, Verlag Chemie, Weinheim 1974, Vol. 4.
[9] D. L. Heron and D. Haneman, *Surface Sci.*, **21**, 12 (1970).
[10] A. E. Morgan and G. A. Samorjai, *Surface Sci.*, **12**, 405 (1968).
[11] P. W. Palmberg and T. N. Rhodin, *Phys. Rev.*, **161**, 586 (1967).
[12] J. T. Grant, *Surface Sci.*, **18**, 228 (1969).
[13] F. Jona, *Surface Sci.*, **8**, 57 (1967).
[14] G. A. Samorjai, *Surface Sci.*, *loc. cit.*, Vol. 1, p. 210.
[15] *Characterization of Solid Surfaces*, Eds. Ph. F. Kane and G. B. Larabee, Plenum Press, New York–London 1974, Chapter 20.
[16] K. Siegbahn, C. Nordling, A. Fahlman, R. Nordberg, K. Hamrin, J. Hedman, G. Johansson, T. Bergmark, S. E. Karlsson, I. Lindgreen and B. Lindberg, *ESCA atomic, molecular and solid state structure studies by means of electron spectroscopy*, Almquist and Niksells, Upsala 1967.
Structure and Bonding, in *Photoelectron Spectroscopy*, Springer Verlag, New York–Berlin–Heidelberg 1975, Vol. 24.
[17] D. A. Eastman and J. K. Cashion, *Phys. Rev. Lett.*, **27**, 1520 (1971).
[18] G. Doyen and G. Ertl, *Surface Sci.*, **43**, 197 (1974).
[19] G. Blyholder, *J. Vac. Sci. Technol.*, **11**, 865 (1974).
[20] J. P. Batra and O. Robaux, *J. Vac. Sci. Technol.*, **12**, 242 (1975).
[21] R. P. Messmer and C. W. Tucker, Jr., *Surface Sci.*, **42**, 341 (1974).
[22] J. E. Demuth and D. E. Eastman, *Phys. Rev. Lett.*, **32**, 1123 (1974).
[23] A. M. Bradshaw, D. Menzel and M. Steinkilberg, *Jap. J. Appl. Phys.*, Suppl. 2, Pt. 2, 841 (1974).
[24] J. P. Page, D. L. Trimm and P. M. Williams, *J. Chem. Soc. Faraday Trans.*, **170,** 1769 (1974).
[25] A. M. Bradshaw, D. Menzel and M. Steinkilberg, *Faraday Discuss. Chem. Soc.*, **58** (1974).
[26] R. Gomer, *Field Emission and Field Ionization*, Harward University Press, Massachusetts 1961.
[27] E. W. Müller, *Phys. Z.*, **37**, 838 (1936); *Z Phys.*, **106**, 541 (1937).
[28] E. W. Müller, *Z. Phys.*, **131**, 136 (1951); *Science*, **149**, 591 (1965).
[29] E. W. Müller and T. T. Tsong, *Field Ion Microscopy*, Elsevier, New York 1969.
[30] K. M. Bowkett and D. A. Smith, *Field Ion Microscopy*, North-Holland Publ. Comp., Amsterdam 1970.
[31] P. A. Redhead, *Can. J. Phys.*, **42**, 886 (1964).
[32] D. Menzel, *Z. Naturforsch.*, **23a**, 330 (1968); *Ber. Bunsenges.*, **72**, 591 (1968).
[33] D. Lichtmann, F. N. Simon and T. R. Kirst, *Surface Sci.*, **9**, 325 (1968).
[34] D. Menzel and R. Gomer, *J. Chem. Phys.*, **41**, 3311 (1964).
[35] D. A. King, *Surface Sci.*, **9**, 375 (1968).
[36] P. A. Redhead, *Vacuum*, **12**, 203 (1962).
[37] H. Simon and R. Suhrmann, *Der lichtelektrische Effekt und seine Anwendungen*, Springer-Verlag, Berlin 1958, p. 26 ff.
[38] P. P. Craig and V. Radeka, *Rev. Sci. Instr.*, **41**, 258 (1970).
[39] A. G. Knapp, *Surface Sci.*, **34**, 289 (1973).
[40] J. Topping, *Proc. Roy. Soc.* (*London*), **A114**, 67 (1972).
[41] E. P. Gyftopulos and J. D. Levine, *J. Appl. Phys.*, **33**, 67 (1962).
[42] N. D. Lang, *J. Vac. Sci. Technol.*, **9**, 569 (1972).

[43] J. Pritchard, *J. Vac. Sci. Technol.*, **9**, 895 (1972).
[44] W. Karpiński and W. Palczewska, *Bull. Acad. Sci. Polon., Ser. Sci. Chim.*, **22**, 159 (1974).
[45] J. Rudny, Ph.D. dissertation, Wrocław 1976.
[46] G. Wedler, *Adsorption*, Verlag Chemie, Weinheim 1970.
[47] W. B. Innes, in *Experimental Methods in Catalytic Research*, Ed. R. B. Anderson, Academic Press, New York 1968, p. 81 ff.
[48] Z. Knor, in *Catalytic Reviews*, Ed. H. Heinemann, Marcel Dekker Inc., New York 1968, Vol. 1, p. 257.
[49] L. H. Little, *Infrared Spectra of Adsorbed Species*, Academic Press, London–New York 1966.
[50] M. L. Hair, *Infrared Spectroscopy in Surface Chemistry*, Marcel Dekker Inc., New York 1967.
[51] R. P. Eischens and W. A. Pliskin, *Adv. in Catalysis*, **10**, 2 (1958).
[52] H. F. Leftin and M. C. Hobson, *Adv. in Catalysis*, **14**, 115 (1963).
[53] A. V. Kiselev and V. I. Lygin, *Usp. Khim.*, **31**, 351 (1962).
[54] W. A. Pliskin and R. P. Eischens, *Z. physik. Chem. (Frankfurt)*, **24**, 11 (1960).
[55] D. D. Eley, D. M. Moran and C. H. Rochester, *Trans. Faraday Soc.*, **64**, 2168 (1968).
[56] R. van Hardeveld and A. van Montfoort, *Surface Sci.*, **4**, 2168 (1968).
[57] P. W. Selwood, *Adsorption and Collective Paramagnetism*, Academic Press, New York–London 1962.
[58] P. W. Selwood, *Chemisorption and Magnetism*, Academic Press, New York–San Francisco–London 1975.
[59] G. A. Martin and B. Imelik, *Surface Sci.*, **42**, 157 (1974).
[60] R. Z. C. van Meerten, T. F. M. de Graaf and J. W. E. Coenen, *J. Catal.*, **46**, 1 (1977).
[61] J. A. Dalmon, C. A. Martin and B. Imelik, *C. R. Acad. Sci. Paris*, **278**, Ser. C, 1481 (1974).
[62] R. Suhrmann, E. A. Dierk, B. Engelke, H. Hermann and K. Schulz, *J. Chim. physiq.*, **1957**, 15.
[63] G. Ertl, M. Neumann and K. M. Strait, *Surface Sci.*, **64**, 393 (1977).
[64] G. Brodén, G. Pirug and H. P. Bonzel, *Surface Sci.*, **72**, 45 (1978).
[65] E. G. M. Kuipers, A. K. Breedijk, W. J. J. van der Waal and J. W. Geus, *J. Catal.*, **72**, 210 (1981).
[66] D. J. C. Yates, L. L. Murell and E. B. Prestridge, *J. Catal.*, **57**, 41 (1979).
[67] J. C. Robertson and C. W. Wilmsen, *J. Vac. Sci. Technol.*, **9**, 901 (1972).
[68] D. J. M. Fassaert, H. Verbeek and A. van Avoird, *Surface Sci.*, **29**, 501 (1972).
[69] L. W. Anders, R. S. Hansen and L. S. Bartell, *J. Chem. Phys.*, **59**, 5277 (1973).
[70] R. Gomer, R. Wortman and R. Lundry, *J. Chem. Phys.*, **26**, 1147 (1957).
[71] L. W. Anders, R. S. Hansen and L. S. Bartell, *J. Chem. Phys.*, **62**, 1641 (1975).
[72] A. B. Anderson and R. Hofmann, *J. Chem. Phys.*, **61**, 4545 (1974).
[73] T. Halachev and E. Ruckenstein, *J. Catal.*, **73**, 171 (1982).
[74] G. Blyholder, *J. Chem. Phys.*, **62**, 3193 (1975).
[75] K. H. Johnson and R. P. Messmer, *J. Vac. Sci. Technol.*, **11**, 236 (1974).
]76] D. E. Eastman and J. Cashion, *Phys. Rev. Letters*, **27**, 1520 (1971).
[77] J. P. Batra and P. Robaux, *Surface Sci.*, **49**, 653 (1975).
[78] J. P. Batra and P. S. Bagus, *Solid State Comm.*, **16**, 1097 (1975).

[79] G. Blyholder, *J. Vac. Sci. Technol.*, **11**, 865 (1974).
[80] J. E. Demuth and D. E. Eastman, *Phys. Rev.*, **B13**, 1523 (1976).
[81] B. N. Cox and C. W. Bauschlicher, Jr., *Surface Sci.*, **108**, 483 (1981), *ibid.*, **102**, 295 (1981).
[82] K. Herrmann and P. S. Bagus, *Phys. Rev.*, **B16**, 4195 (1977).
[83] D. E. Ellis, E. J. Barends, H. Adachi and F. W. Averill, *Surface Sci.*, **64**, 649 (1977).
[84] G. C. Bond, *Heterogeneous Catalysis: Principles and Applications*, Clarendon Press, Oxford 1974.
[85] J. M. Thomas and W. J. Thomas, *Introduction to Principles of Heterogeneous Catalysis*, Academic Press, New York 1967.
[86] A. Clark, *Theory of Adsorption and Catalysis*, Academic Press, New York 1970.
[87] M. Boudart *Heterogeneous Catalysis*, in *Physical Chemistry*, Eds. H. Eyring, D. Henderson and W. Jost, Academic Press, New York 1975, Vol. 7, Chapter 7.
[88] A. A. Balandin, *Z. physik. Chem.*, **B2**, 289 (1929); **B3**, 167 (1929).
[89] N. I. Kobozev, *Usp. Khim.*, **25**, 545 (1956).
[90] M. Boudart, *Adv. Catal.*, **20**, 153 (1969).
[91] J. Mc Allister and R. S. Hansen, *J. Chem. Phys.*, **59**, 414 (1973).
[92] D. J. C. Yates and J. H. Sinfelt, *J. Catal.*, **8**, 348 (1967).
[93] J. R. Anderson and Y. Schimoyama, *Proc. V-th Intern. Congr. Catal.*, Ed. J. Hightower, North-Holland Publ. Comp., Amsterdam 1973, p. 695.
[94] J. H. Sinfelt, J. L. Carter and D. J. C. Yates, *J. Catal.*, **24**, 283 (1972).
[95] S. Fuentes, F. Figureas and R. Gomez, *J. Catal.*, **68**, 419 (1981).
[96] G. L. Ertl and J. Koch., *Proc. V-th Intern. Congr. Catal.*, *loc. cit.*, p. 969.
[97] M. Boudart, A. W. Aldag, J. E. Benson, N. A. Dougharty and C. G. Harkins, *J. Catal.*, **6**, 92 (1966).
[98] M. Boudart, in *Interaction on Metal Surfaces*, Springer-Verlag, New Nork 1975, p. 275.
[99] R. van Hardeveld and F. Hartog, *Surface Sci.*, **15**, 189 (1969).
[100] L. Pauling, *Proc. Roy. Soc. (London)*, **A196**, 343 (1949).
[101] O. Beeck, *Discuss. Faraday Soc.*, **8**, 118 (1950).
[102] G. C. Bond, *Catalysis by Metals*, Academic Press, London 1962; *Discuss. Faraday Soc.*, **41**, 200 (1966).
[103] J. A. Cusumano, R. A. Dalla Betta and R. B. Levy, *Catalysis in Coal Conversion*, Academic Press, New York–San Francisco–London 1978.
[104] P. van der Plank and W. M. H. Sachtler, *J. Catal.*, **7**, 300 (1967).
[105] J. H. Sinfelt, J. C. Carter and D. J. C. Yates, *J. Catal.*, **24**, 293 (1972).
[106] W. Trzebiatowski and H. Kubicka, *Z. Chem.*, **3**, 262 (1963).
[107] W. Romanowski and J. Rudny, *Roczniki Chem.*, **41**, 1221 (1967).
[108] R. Bouwman, G. I. Lippits and W. M. H. Sachtler, *J. Catal.*, **25**, 350 (1972).
[109] B. J. Wood and H. Wise, *Surface Sci.*, **52**, 151 (1975).
[110] F. L. Williams and M. Boudart, *J. Catal.*, **30**, 438 (1973).
[111] R. A. van Santen and M. A. M. Boersma, *J. Catal.*, **34**, 13 (1974).
[112] F. L. Williams and D. Nanson, *Surface Sci.*, **45**, 377 (1974).
[113] R. L. Park and J. E. Houston, *J. Vac. Sci. Technol.*, **10**, 176 (1973).
[114] J. J. Burton, E. Hyman and D. G. Fedak, *J. Catal.*, **37** 106 (1975).
[115] J. H. Sinfelt, *J. Catal.*, **29**, 308 (1973).
[116] J. H. Sinfelt, *Catal. Rev. Sci. Eng.*, **9**, 147 (1974).

[117] E. L. Muetterties, *Science*, **196**, 839 (1977).
[118] E. L. Muetterties, T. N. Rhodin, E. Baud, C. F. Brucker and W. R. Pretzer, *Chem Rev.*, **79**, 91 (1979).
[119] J. W. Lauher, *J. Am. Chem. Soc.*, **100**, 5305 (1978).
[120] M. Moskovits, *Acc. Chem. Res.*, **12**, 229 (1979).
[121] P. Chini, G. Longoni and V. G. Albano, *Adv. Organomet. Chem.*, **14**, 285 (1976).
[122] E. L. Muetterties, *Pure Appl. Chem.*, **50**, 941 (1978).
[123] G. C. Demitras and E. L. Muetterties, *J. Am. Chem. Soc.*, **99**, 2796 (1977).
[124] A. K. Smith, A. Theolier, J. M. Basset, R. Ugo, D. Commereuc and Y. Chauvin, *J. Am. Chem. Soc.*, **100**, 2590 (1978).
[125] R. M. Laine, R. G. Rinker and P. C. Ford, *J. Am. Chem. Soc.*, **99**, 252 (1977).
[126] G. C. Smith, T. P. Chojnacki, S. R. Dasgupta and K. L. Watters, *Inorg. Chem.*, **14**, 1419 (1975).
[127] K. L. Watters, R. F. Howe, T. P. Chojnacki, Ch. W. Fu, R. L. Schneider and N. B. Wong, *J. Catal.*, **66**, 424 (1980).
[128] A. K. Smith, F. Hugues, A. Theolier, J. M. Basset, R. Ugo, G. M. Zanderghini, J. L. Bilhou-Bougnol and W. F. Graydon, *Inorg. Chem.*, **18**, 3104 (1979).
[129] V. L. Kuznetsov, A. T. Bell and Y. I. Yermakov, *J. Catal.*, **65**, 374 (1980)
[130] M. Deeba and B. C. Gates, *J. Catal.*, **67**, 303 (1981).
[131] H. Knözinger and Y. Zhao, *J. Catal.*, **71**, 337 (1981).
[132] B. C. Gates and J. Lieto, *Chem. Technol.*, **195**, 248 (1980).
[133] M. B. Freeman, M. A. Patrick and B. C. Gates, *J. Catal.*, **73**, 82 (1982).
[134] F. Hughues, A. K. Smith, Y. Ben Taarit, J. M. Basset, D. Commereuc and Y. Chauvin, *J. Chem. Soc. Chem. Commun.*, **1980**, 68.
[135] M. Ichikawa, *Bull. Soc. Chem. Japan*, **51**, 2268 (1978).

Additional References

Chapter 1

Log-normal distribution of metal particle sizes

W. I. Kvashonkin and D. A. Agievskii, *Zavodsk. Laborat.*, **45**, 730 (1979).
S. K. Kurtz and F. M. A. Carpay, *J. Appl. Phys.*, **51**, 5725 (1980).
M. A. Kipnis and D. A. Agievskii, *Kinet. Kataliz*, **23**, 236 (1982).
D. A. Agievskii and W. I. Kvashonkin, *ibid.*, **25**, 218 (1984).

Chapter 2

Raney metals

P. Fouilloux, *Appl. Catal.*, **8**, 1 (1983).
J. P. Orchard, A. D. Tomsett, M. S. Wainwright and D. J. Young, *J. Catal.*, **84**, 189 (1983).

Matrix isolation of metal clusters

S. C. Davis and K. J. Klabunde, *Chem. Rev.*, **82**, 153 (1982).
M. Seidel and M. Grosch, *Wiss. Z. Friedrich Schiller Univ. Jena, Math. Naturwiss. Reihe*, **31**, 1065 (1982).

Gas evaporation

Y. Saito, K. Mihama and R. Uyeda, *Jap. J. Appl. Phys.*, **19**, 1603 (1980).
M. Kobayashi, S. Yatsuya and K. Mihama, *ibid.*, **20**, 805 (1981).
W. Schulze, F. Frank, K. P. Charlé and B. Tesche, *Ber. Bunsenges. physik. Chem.*, **88**, 363 (1984).
S. Iwama and K. Hayakawa, *Surface Sci.*, **156** (p. 1), 85 (1985).
F. Frank, W. Schulze, B. Tesche, J. Urban and B. Winter, *ibid.*, **156** (p. 1), 90 (1985).

Multiple expansion cluster source (ultrasonic jet) — technique

R. S. Bowles, S. B. Park and R. P. Andres, *J. Mol. Catal.*, **20**, 279 (1983).

Ultrasonic jet technique

M. D. Morse and R. E. Smalley, *Ber. Bunsenges. physik. Chem.*, **88**, 228 (1984).
J. Gspan, *ibid.*, **88**, 256 (1984).
G. Delacrétáz and L. Wöste, *Surface Sci.*, **156** (p. 2), 770 (1985).

Aqueous sols

I. N. Furlong, A. Launikonis, W. H. F. Sasse and J. V. Sanders, *J. Chem. Soc. Faraday Trans.*, *I* **80**, 803 (1984).
D. Schönauer and U. Kreibig, *Surface Sci.*, **156** (p. 1), 100 (1985).

Chapter 3

Magnetization

G. A. Martin, R. Dutartre and J. A. Dalmon, in *Growth and Properties of Metal Clusters*, Ed. J. Bourdon, Elsevier, Amsterdam 1980, p. 467.
W. Romanowski, *Polish J. Chem.*, **54**, 1515 (1980).
S. Engels, W. Mörke, M. Wilde, W. Roschke, B. Freitag and H. Siegel, *Z. anorg. allg. Chem.*, **472**, 162 (1981).
F. S. Cale and J. T. Richardson, *J. Catal.*, **79**, 378 (1983).
F. Schmidt, U. Stapel and H. Walther, *Ber. Bunsenges. physik. Chem.*, **88**, 310 (1984).
T. S. Cale and D. K. Ludlow, *J. Catal.*, **86**, 450 (1984).
K. Kimura, S. Bandow and S. Sako, *Surface Sci.*, **156** (p. 2), 883 (1985).

Optical properties

K. H. Bennemann and S. Reindl, *Ber. Bunsenges. physik. Chem.*, **88**, 278 (1984).
U. Kreibig and L. Genzel, *Surface Sci.*, **156** (p. 2), 678 (1985).
M. Quinten, U. Kreibig, D. Schonauer and L. Genzel, *ibid.*, **156** (p. 2), 741 (1985).
J. E. Inglesfield, *ibid.*, **156** (p. 2), 830 (1985).
A. Ljungbert and S. Lundquist, *ibid.*, **156** (p. 2), 839 (1985).
F. Cocchini, F. Basani and M. Bourg, *ibid.*, **156** (p. 2), 851 (1985).

X-ray scattering and EXAFS

R. W. Joyner, *J. Chem. Soc. Faraday Trans. I*, **76**, 357 (1980).
G. Cocco, L. Schiffini, G. Strukul and G. Carturan, *J. Catal.*, **65**, 348 (1980).
D. G. Mustard and C. H. Bartolomew, *ibid.*, **67**, 186 (1981).

B. Moraweck and A. J. Renouprez, *Surface Sci.*, **106**, 35 (1981).
J. H. Sinfelt, G. H. Via and F. W. Lyttle, *J. Chem. Phys.*, **76**, 2779 (1982).
R. K. Nandi, P. Georgopoulos, J. B. Cohen, J. B. Butt, R. L. Burwell, Jr. and D. H. Bilderbock, *J. Catal.*, **77**, 421 (1982).
R. K. Nandi, F. Molinaro, C. Tang, J. B. Cohen, J. B. Butt and R. L. Burwell, Jr., *ibid.*, **78**, 289 (1982).
M. Ito and H. Iwasaki, *Jap. J. Appl. Phys.*, **22**, 357 (1983).
G. H. Via, G. Meitzner, F. W. Lyttle and J. H. Sinfelt, *J. Phys. Chem.*, **79**, 1527 (1983).

Chapter 4

General, semi-empirical quantum chemistry methods

R. C. Baetzold, M. G. Mason and J. F. Hamilton, *J. Chem. Phys.*, **72**, 366 (1980).
H. Müller and Ch. Opitz, *Z. phys. Chem.* (*Leipzig*), **261**, 761 (1980).
T. Halicioglu and P. J. White, *Surface Sci.*, **106**, 45 (1981).
R. P. Messmer, *ibid.*, **106**, 225 (1981).
R. C. Baetzold, *ibid.*, **106**, 243 (1981).
D. R. Salahub and R. P. Messmer, *ibid.*, **106**, 415 (1981).
G. A. Thomson and D. M. Lindsay, *J. Chem. Phys.*, **74**, 959 (1981).
J. Oksuz, *Surface Sci.*, **122**, L582 (1982).
T. Yanagihara, *Jap. Journ. Appl. Phys.*, **21**, 1554 (1982).
J. H. Mühlbach, K. Sattler, P. Pfau and E. Recknagel, *Phys. Lett.*, **87A**, 418 (1982).
R. P. Messmer and S. H. Lamson, *Chem. Phys. Lett.*, **90**, 31 (1982).
M. A. Listvan, *J. Mol. Catal.*, **20**, 265 (1983).
G. Mason, *Phys. Rev.*, **B27**, 748 (1983).
B. Delley, D. E. Ellis, A. J. Freeman, E. J. Baerends and D. Pot, *ibid.*, **27**, 2132 (1983).
D. Tománek, S. Muherjee and K. H. Bennemann, *ibid.*, **28**, 665 (1983).
W. B. Moonsmith and J. M. Cowley, *Ultramicroscopy*, **12**, 177 (1984).
Sheng-Wei Wang, L. M. Falicov and A. V. Scarey, *Surface Sci.*, **143**, 609 (1984).
R. L. Whetten, D. M. Cox, D. J. Trevor and A. Kaldor, *Surface Sci.*, **156**, 8 (1985).
L. R. Wallenberg, J.-O. Bovin and G. Schmid, *ibid.*, **156** (p. 1), 256 (1985).
E. Blaisten-Barojas and H. C. Andersen, *ibid.*, **156** (p. 1), 548 (1985).
T. Halicioglu, T. Takai and W. A. Tiller, *ibid.*, **156** (p. 1), 556 (1985).
J. P. Borel and A. Châtelain, *ibid.*, **156** (p. 2), 572 (1985).
P. G. Mezey, *ibid.*, **156** (p. 2), 597 (1985).

Ab initio calculations

C. Bachmann, J. Demuynck and A. Veillard, in *Growth and Properties of Metal Clusters*, *loc. cit.*, p. 269.
H. O. Beckmann, J. Koutecký and V. Bonačić-Koutecký, *J. Chem. Phys.*, **73**, 5182 (1980).
J. Demuynck, M. M. Rohmer, A. Strich and A. Veillard, *ibid.*, **75**, 3443 (1981).

H. T. Tatewaki, E. Miyoshi and T. Nakamura, *ibid.*, **76**, 5073 (1982); *ibid.* **78**, 815 (1983).
E. Miyoshi, *Intern. J. Quant. Chem.*, **23**, 1201 (1983).
G. H. Jeung, M. Pelissier and J. C. Barthelat, *Chem. Phys. Lett.*, **97**, 369 (1983).
J. Koutecky and G. Pacchioni, *Ber. Bunsenges. physik. Chem.*, **88**, 233 (1984).
G. Pacchioni and J. Koutecky, *ibid.*, **88**, 242 (1984).
K. A. Gingerich, I. Shim, S. K. Gupta and J. E. Kingcade, Jr., *Surface Sci.*, **156** (p. 1), 495 (1985).
P. S. Bagus, C. J. Nelin and C. W. Bauschlicher, Jr., *ibid.*, **156** (p. 2), 615 (1985).
J. Koutecky, G. Pacchioni, G. H. Jeung and E. C. Hass, *ibid.*, **156** (p. 2), 650 (1985).

Chapter 5

Surface tension

M. Fortes, *J. Chem. Soc. Faraday Trans. I*, **78**, 101 (1982).
V. K. Kumikov and Kh. B. Khokonov, *J. Appl. Phys.*, **54**, 1346 (1983).

Metal–support interaction

M. A. Vannice and R. L. Garten, *J. Catal.*, **63**, 255 (1980).
R. B. Gregor and F. W. Lyttle, *ibid.*, **63**, 476 (1980).
R. T. K. Baker, *ibid.*, **63**, 523 (1980).
M. J. Yacamán and J. M. Dominguez, *ibid.*, **64**, 213 (1980).
J. M. Dominguez and M. J. Yacamán, *ibid.*, **64**, 223 (1980).
X. E. Verykios, F. P. Stein and R. W. Coughlin, *ibid.*, **66**, 147 (1980).
C. W. Keep, S. Terry and M. Wells, *ibid.*, **66**, 451 (1980).
H. Praliaud and G. A. Martin, *ibid.*, **72**, 394 (1981).
G. A. Martin and J. A. Dalmon, *React. Kin. Catal. Lett.*, **16**, 325 (1981).
G. A. Martin, R. Dutartre and J. A. Dalmon, *ibid.*, **16**, 329 (1981).
S. C. Fung, *J. Catal.*, **76**, 225 (1982).
B. A. Sexton, A. E. Hughes and K. Foger, *ibid.*, **77**, 85 (1982).
P. Turlier and G. A. Martin, *React. Kin. Catal. Lett.*, **21** 387 (1982).
G. C. Bond and R. Burch, *Spec. Period. Rep. Catalysis*, Vol. 6, The Roy. Soc. Chem., London 1983, p. 27.
K. Kunimori, N. Ikeda, M. Soma and T. Uchijima, *J. Catal.*, **79**, 185 (1983).
R. T. K. Baker, E. B. Prestridge and L. L. Murell, *ibid.*, **79**, 348 (1983).
C. H. Bartholomew and W. L. Sorensen, *ibid.*, **81**, 131 (1983).
M. J. Kelley, D. R. Short and D. G. Schwartzfager, *J. Mol. Catal.*, **20**, 235 (1983).
D. E. Resasco and G. L. Haller, *Appl. Catal.*, **8**, 99 (1983).
A. G. Shastri, A. K. Datye and J. Schwank, *J. Catal.*, **87**, 265 (1984).
Li Wenzhao, Chen Yixuan, Yu Chunying, Wang Xiangzhen. Hong Zupei and Wei Zhaobin, *Proc. 8th Intern. Congr. Catal.*, Verlag Chemie, Berlin 1984, Weinheim 1984, Vol. 5, p. 205.

K. Kunimori, H. Abe, E. Yamaguchi, S. Matsui and T. Uchijima, *ibid.*, Vol. 5, p. 251.
J. M. Herrmann, *J. Catal.*, **89**, 404 (1984).
H. Takatani and Y. W. Chung, *ibid.*, **90**, 75 (1984).
W. Romanowski and R. Lamber, *Thin Solid Films*, **127**, 139 (1985).
R. Lamber, *ibid.*, **128**, L29 (1985).

Surface diffusion and aggregation of metals

S. Iwama and T. Sahashi, *Jap. J. Appl. Phys.*, **19**, 1039 (1980).
T. K. Galeev, N. N. Bulgakov, G. A. Savelieva and N. M. Popova, *React. Kin. Catal. Lett.*, **14**, 61 (1980).
P. Wynblatt and J. K. Tien, *J. Catal.*, **62**, 59 (1980).
H. H. Lee, *ibid.*, **62**, 129 (1980).
H. K. Kuo, P. Ganesan and J. R. Angelis, *ibid.*, **64**, 303 (1980).
R. T. K. Baker, R. D. Sherwood and J. A. Dumesic, *ibid.*, **66**, 56 (1980).
T. Wang and L. D. Schmidt, *ibid.*, **66**, 301 (1980).
T. M. Ahn, J. K. Tien and P. Wynblatt, *ibid.*, **66**, 335 (1980).
P. F. Flahive and W. R. Graham, *Surface Sci.*, **91**, 449 (1980).
R. T. Tung and W. R. Graham, *ibid.*, **97**, 73 (1980).
G. Ehrlich and K. Stolt, in *Growth and Properties of Metal Clusters*, *loc. cit.*, p. 1.
P. Wynblatt, *ibid.*, p. 15.
S. Iwama and K. Hayakawa, *Jap. J. Appl. Phys.*, **20**, 335 (1981).
U. M. Titulaer and J. M. Deutch, *J. Chem. Phys.*, **77**, 472 (1982).
J. D. Doll and H. K. Mc Dowell, *ibid.*, **77**, 479 (1982).
M. Arai, T. Ishikawa and Y. Nishiyama, *J. Phys. Chem.*, **86**, 577 (1982).
R. Gomer, *Vacuum*, **33**, 537 (1983).
E. Ruckenstein and S. H. Lee, *J. Catal.*, **86**, 457 (1984).
T. Nakayama, M. Arai and Y. Nishiyama, *ibid.*, **87**, 108 (1984).

Nucleation

J. A. Venables, *Vacuum*, **33**, 701 (1983).

Chapter 6

Structure and composition of metal surfaces

T. T. Tsong, Yee S. Ng and S. B. Mc Lane, Jr., *J. Chem. Phys.*, **73**, 1464 (1980).
K. Christmann, G. Ertl and H. Shimizu, *J. Catal.*, **61**, 397 (1980).
P. Biloen, J. N. Helle, H. Verbeed, F. M. Dautzenberg and W. M. H. Sachtler, *ibid.*, **63**, 112 (1980).
H. C. de Jongste and V. Ponec, *ibid.*, **64**, 228 (1980).
S. Anderson and J. B. Pendry, *J. Phys. C* (*Solid State Phys.*), **13**, 3547 (1980).

H. B. Nielsen and D. L. Adams, *Surface Sci.*, **97**, L351 (1980).
J. P. Biberian, *ibid.*, **97**, 257 (1980).
C. A. Balseiro and J. L. Morán-Lopez, *Phys. Rev.*, **B21**, 349 (1980).
D. J. Chadi, R. S. Bauer, R. H. Williams, G. V. Hansen, R. Z. Bechrach J. M. Mikkelsen, F. Honzay, G. M. Guichard, R. Pinchaux and Y. Petroff, *Phys. Rev. Lett.*, **44**, 799 (1980).
M. P. Sheah, *J. Phys. F* (*Metal Phys.*), **10**, 1043 (1980).
T. T. Tsong, Yee S. Ng and S. B. Mc Lane, Jr., *J. Appl. Phys.*, **51**, 6189 (1980).
M. A. Van Hove, R. J. Koestner, P. C. Stair, J. B. Biberian, L. L. Kesmodel, I. Bartos and G. A. Samorjai, *Surface Sci.*, **103**, 189, 218 (1981).
F. F. Abraham, *Phys. Rev. Lett.*, **46**, 546 (1981).
K. Hoshimo, *J. Phys. Soc. Japan*, **50**, 577 (1981).
R. K. Wild, *Vacuum*, **31**, 183 (1981).
J. Pollman, *Phys. Rev. Lett.*, **49**, 1649 (1982).
S. Ferrer and H. P. Bonzel, *Surface Sci.*, **119**, 234 (1982).
D. P. Woodruff, *ibid.*, **122**, L653 (1982).

Adsorption on metal surfaces (*experimental work*)

A. Renouprez, P. Fouilloux and J. P. Candy, in *Growth and Properties of Metal Clusters*, *loc. cit.*, p. 537.
K. Kandasamy and N. A. Surplice, *J. Phys. C* (*Solid State Phys.*), **13**, 689 (1980).
H. Shimizu, K. Christmann and G. Ertl, *J. Catal.*, **61**, 412 (1980).
F. C. J. M. Toolemaar, D. Reinalda and V. Ponec, *ibid.*, **64**, 110 1(980).
J. G. Goodwin, Jr. and C. Naccache, *ibid.*, **64**, 482 (1980).
Ch. M. Grill and R. D. Gonzales, *ibid.*, **64**, 487 (1980).
G. B. Mc Vicker, R. T. K. Baker, R. L. Garten and E. R. Kügler, *ibid.*, **65**, 207 (1980).
A. M. Baró and H. Ibach, *Surface Sci.*, **92**, 237 (1980).
C. W. Seabury, T. N. Rhodin, R. J. Purtell and R. P. Merill, *ibid.*, **93**, 117 (1980).
D. W. Goodman, J. T. Yates and T. E. Madey, *ibid.*, **93**, L135 (1980).
H. Niehus and G. Comsa, *ibid.*, **93**, L147 (1980).
J. L. Gland, *ibid.*, **93**, 487 (1980).
J. S. Ford, P. J. Goddard and R. M. Lambert, *ibid.*, **94**, 339 (1980).
M. E. Bridge and R. M. Lambert, *ibid.*, **94**, 469 (1980).
M. A. Barteau and R. J. Madix, *ibid.*, **97**, 101 (1980).
F. H. P. M. Habraken, C. M. A. M. Mesters and G. A. Bootsma, *ibid.*, **97**, 264 (1980).
S. Engels, Nguyen Phuong Khue and M. Wilde, *Z. anorg. allg. Chem.*, **461**, 155 (1980).
K. Jacobi, E. S. Jensen, T. N. Rhodin and R. P. Merill, *Surface Sci.*, **108**, 205 (1981).
R. G. Greenler, D. R. Snider, D. Witt and R. S. Sorbello, *ibid.*, **118**, 415 (1982).
J. P. Biberian and M. A. Van Hove, *ibid.*, **118**, 443 (1982).
K. Horn, J. Di Nardo, W. Eberhardt, M. J. Freund and E. W. Plummer, *ibid.*, **118**, 465 (1982).
A. Ortega, F. M. Hoffman and A. M. Bradshaw, *ibid.*, **119**, 79 (1982).
F. P. Netzer and T. E. Madey, *ibid.*, **119**, 422 (1982).
G. Popov and E. Bauer, *ibid.*, **122**, 433 (1982).
St. A. Astegger and E. Bechtold, *ibid.*, **122**, 491 (1982).
D. L. Doering, J. T. Dickinson and H. Poppa, *J. Catal.*, **73**, 91 (1982).

Chau Hwa Yang and J. J. Goodwin, Jr., *React. Kin. Catal. Lett.*, **20**, 13 (1982).
E. G. M. Kuijpers, A. K. Breedijk, W. J. J. van der Wal and J. W. Geus, *J. Catal.*, **72**, 210 (1981); *ibid.*, **81**, 429 (1983).
E. S. Jensen and T. N. Rhodin, *Phys. Rev.*, **B27**, 3338 (1983).
D. Schmeisser, F. J. Himpsel and G. Hollinger, *ibid.*, **B27**, 7813 (1983).
M. I. Shek, P. M. Stefan, I. Lindau and W. E. Spicer, *ibid.*, **B27**, 7277 (1983); *ibid.*, **B27**, 7301 (1983).
K. H. Rieder, *ibid.*, **B27**, 7799 (1983).
J. Bénard (Ed.), *Adsorption on Metal Surfaces — An Integrated Approach*, Elsevier, Amsterdam 1983.
E. Bornand, Surface Sci., **152/153**, 314 (1985).

Adsorption, chemisorption (theory including quantum chemical methods)

A. B. Meador, Te-Hua Lin and H. B. Huttington, *Surface Sci.*, **97**, 53 (1980).
K. A. R. Mitchell, *ibid.*, **100**, 225 (1980).
J. L. Whitten and T. A. Pakkanen, *Phys. Rev.*, **B21**, 4357 (1980)
C. M. Varma and A. J. Wilson, *ibid.*, **B22**, 3795 (1980).
C. M. Varma and W. Andreoni, *ibid.*, **B23**, 437 (1980).
J. Lopez, J. C. le Bosee and J. Rousseau-Violett, *J. Phys. C (Solid State Phys.)*, **13**, 1139 (1980).
H. H. Dunken and E. Jemmis, *Z. Chem.*, **20**, 454 (1980).
H. Kasai and A. Okiji, *J. Phys. Soc. Japan*, **50**, 2363 (1981).
J. L. Morán-Lopez and L. M. Falicov, *Phys. Rev.*, **B26**, 2560 (1982).
J. Paul and A. Rosén, *ibid.*, **B26**, 4073 (1982).
C. R. Fisher, L. A. Burke and J. L. Whitten, *Phys. Rev. Lett.*, **49**, 344 (1982).
J. Küppers and U. Seip, *Surface Sci.*, **119**, 291 (1982).
H. Adachi, M. Tsukada, I. Yasumori and M. Onchi, *ibid.*, **119**, 10 (1982).
E. Ruckenstein and T. Halachev, *ibid.*, **122**, 422 (1982).
W. Eberhardt, S. G. Louie and E. W. Plummer, *Phys. Rev.*, **B28**, 465 (1983).
G. Vidali, M. W. Cole and J. R. Klein, *ibid.*, **B28**, 3064 (1983).
E. Wimmer, A. J. Freeman, J. R. Hiskes and A. M. Karo, *ibid.*, **B28**, 3074 (1983).
R. A. van Santen, *Proc. 8th Intern. Congr. Catal.*, Verlag Chemie Berlin 1984, Weinheim 1984, Vol. 4, p. 97.
G. Abbate, V. Barone, F. Lelj, E. Iaconis and N. Russo, *Surface Sci.*, **152/153**, 690 (1985).
D. W. Bullet, *ibid.*, **152/153**, 654 (1985).
E. Miyoshi, Y. Sakai and S. Mori, *ibid.*, **156** (p. 2), 667 (1985).
F. Raatz and D. R. Salahub, *ibid.*, **156** (p. 2), 892 (1985).

Catalysis by molecular metal clusters

B. F. G. Johnson (Ed.)., *Transition Metal Clusters*, Wiley, New York 1980.
E. L. Muetterties, *Chem. Eng. News*, 1982, 28.
E. L. Muetterties and M. J. Krause, *Angew. Chem. Int. Ed.*, **22**, 135 (1983).
T. H. Maugh II, *Science*, **220**, 592 (1983).

Electronic factor in catalysis

J. K. Norskov, *Proc. 8th Intern. Congr. Catal.*, Verlag Chemie, Berlin 1984, Weinheim 1984, Vol. 4, p. 85.

Geometric factor in catalysis

D. W. Goodman, *Proc. 8th Intern. Congr. Catal.*, Verlag Chemie, Berlin 1984, Weinheim 1984, Vol. 4, p. 3.

General

G. A. Samorjai, *Proc. 8th Intern. Congr. Catal.*, Verlag Chemie, Berlin 1984, Weinheim 1984, Vol. 1, p. 113.

G. C. Bond, *Surface Sci.*, **156** (p. 2), 966 (1985).

Index

ab initio quantum chemistry methods 109, 110, 181
abrasion 122
absorption of light 73, 76
  X-days 83ff
activation, energy of electrical conduction 65
  reduction 33
    surface diffusion 136
activity, catalytic 182
active centres in catalysis 181, 185
  complex 181
Adams method to obtain metal "blacks" 34, 140
adhesion 120 ff
adsorption 30, 46, 166 ff
  isobar 167
  isostere 167, 169
adsorption isotherm 166 ff
  BET 168
  Dubinin–Radushkevich 168, 169
  Freundlich 167
  Langmuir 167
aggregates, of particles 31
alloys 13, 39
  as catalysts 186 ff
  Ag–Pd 102, 187
  Cu–Ni 39, 40, 102
  Fe–Co 38
  Fe–Ti 125
  Ni–Au 187
  Ni–Pd 40, 102
  Ni–Re 40
  Ni–Rh 40
  Pd–Co 40
  Pd–Rh 40
  Pd–Ru 40
  Pt–Co 40
  Pt–Fe 40
  Pt–Ir 40, 56
  Pt–Ni 40
  Pt–Pd 40
  Pt–Rh 40, 186
  Pt–Ru 40
  of gold with Cd, Cu, Fe, Ni, Pb, Zn 62
  on carries, Cu–Ni 56
    Cu–Os 56
    Cu–Ru 56
    Pt–Cu 56
    Pt–Fe 56
    Pt–Re 56
antimony 39, 150
"atomization" 41
Auger electron spectroscopy (AES) 151ff, 175, 187
A-15 structure 98

barium 12
binding energy per atom 100 ff
bismuth 150
Born–Oppenheimer approximation 106
Bremsstrahlung 85

broadening of X-ray lines 77
Brownian movements of particles 74, 128

cadmium 12, 41, 101, 122, 123
capture radius 46
carbon monoxide as reducing agent 31
Cauchy, diffraction line profile 78 ff
centre of the distribution 18
Chebyshev theorem 19
chemisorption, heat of 146
chlorides of metals, reduction of 32
chloroplatinic acid 34, 36, 37
chromium 34, 35, 66, 98, 123
coalescence of particles 130 ff, 143
cobalt 13, 15, 31, 37, 54, 97, 98
co-condensation 49
cohesion 12
collective paramagnetics 69
"colloidal" metals 12, 36, 41, 48
complete neglect differential orbital (CNDO) 100, 102 ff, 112, 180
conduction, electrical of island films 64 ff
conductivity, electrical 61, 174
configuration interaction (CI) method 109
contact angle 120, 123
copper 13, 15, 31, 37, 41, 62, 85, 97, 98, 102, 122, 171, 179, 186

Coulomb operator 107
  integral 108
Curie temperature, of small particles 68

d-character 185
decomposition, thermal 31, 38
demanding reactions 184
desorption 46
  electron beam-induced 148, 158 ff
  heat of 139
  spectrum of 159, 160
  thermal 148, 158 ff, 175
  thermal activation energy of 159
diffusion of particles (surface) 129
dimension of a particle 16
dipole moment of adsorbed molecules 165
dispersion 14, 15, 19
distribution function 17, 71
  beta 21
  Cauchy 20
  cumulative 18
  Davies 22
  gamma 21
  Gaussian (normal) 20, 24
  general properties of 18
  lognormal 24 ff
  Weibull 23
distribution parameters 18, 82
domains, coherently scattering 30
  magnetic 67

electrolysis 30, 38
electron affinity (EA) 100, 112
  emission Schottky 64
  emission thermal 64, 162
  energy levels 99
  microscopy, high resolution 88, 97, 125
  spectroscopy for chemical analysis (ESCA) 148, 153 ff
electronic factor in catalysis 183, 185 ff
energy operator 107
entropy, of clusters 96, 97
error function 27
evaporation, in an electric arc 40, 41
  in an inert atmosphere 41, 42, 44
  in vacuum 41, 45, 47
exchange integral 108
  operator 107
expectation value 18
extended Hueckel molecular orbital (EHMO) method 100–104, 112, 176 ff
  X-ray absorption fine structure (EXAFS) 83 ff
extinction coefficient 76

facile reactions 184
F-centres 48, 49
Fermi energy 99
ferromagnetic small particles 67 ff
field desorption 157

emission microscope 155
emission of electrons 97, 148, 155 ff
films, metallic 45 ff
Fischer–Tropsch synthesis 33, 54, 190, 191
Fock operator 107
forms of adsorption 161
of CO on metals 171
of ethylene on Ni 172, 174
of H on Pt 171
Fourier analysis of X-ray line intensity 79
Fowler equation 162
Fowler–Nordheim equation 163
free energy of atom clusters 93, 96

gallium 12
Gaussian orbitals 117
geometric factor in catalysis 183 ff
germanium 127, 150
gold 13, 15, 36, 37, 48, 62, 65, 66, 76, 97, 98, 102, 122, 123, 150, 163
Guinier-Preston zones 82
Guinier radius 80
Gyftopulos–Levine formula 166

Hamiltonian 106, 109
matrix elements 108, 112
Hartree–Fock–Rothaan equation 108
model 107
heat capacity 96 ff
of adsorption 46
of adsorption, isosteric 169
heterogeneity of surfaces 160, 161
highest occupied molecular orbital (HOMO) 100 ff
histogram 17, 18
Hueckel method 115
hydrides of metals 35, 39
hydrogen, reduction by 31, 32, 34, 36–38, 40, 53–56, 97
hydrogenation 31, 34, 51
hydrometallurgy 13, 38

icosahedral clusters 92, 190
implantation of ions 51
impregnation 52
indium 12
induction period of reduction 33
infrared spectrometry of adsorbed molecules 170 ff
intermediate neglect of differential overlap (INDO) method 111
ion exchange 53
ion field microscope 157 ff
ionisation potential (IP) 100, 103
iridium 13, 15, 34, 55, 150
iron 15, 31, 33, 34, 38, 54, 62, 121, 123, 171
island films 63, 65
iterative extended Hueckel method (IEHM) 113

Kratky, X-ray diffraction camera 83
kurtosis 20

Lambert law 74
Langevin function 69
lattice polyhedra 89
reciprocal 149
lead 39, 123
Lennard–Jones potential 91
linear combination of atomic orbitals (LCAO) method 106, 107, 109, 117
linear combination of atomic orbitals, self consistent field (LCAO–SCF) method 109, 181
lithium 48
location parameters of Weibull distribution 23
London forces 122
low energy electron diffraction (LEED) 149 ff
lowest unoccupied molecular orbital (LUMO) 100 ff

magnesium 12, 123
magnetization of small particles 69, 71, 126
change of by adsorption 172 ff
materials, magnetic 31, 38
matrix isolation 49 ff

mean, arithmetic 16, 18, 28
geometric 16, 28
harmonic 16
relaxation time 139
residence time 139
mechanical methods of comminution 51
median 19
mercury 12, 38
metal-ceramics adhesion 121
metallic molecular clusters 188 ff
mode of the distribution function 19
modified intermediate neglect of differential overlap (MINDO) method 111
molecular beam 45
orbital 106
molybdenum 13, 15, 98, 123
moment, magnetic 67, 68
moments of the distribution function 19, 27
lognormal 27 ff
monodisperse systems 15
multiple scattering method 116

neglect of diatomic differential overlap (NDDO) method 110
nickel 13, 15, 31, 37, 38, 41, 50, 51, 54, 55, 62, 63, 65, 70, 98, 103, 104, 121, 123, 125, 126, 155, 171–173, 176–178
normal distribution,se e Gaussian distribution
nucleation 33, 46, 138 ff
nuclei of metallic phase 33, 39, 46, 140
critical 141

ordered adsorbed layers 151
osmium 13
Ostwald ripening 133 ff
oxalates of metals 31
oxides of metals, reduction of 32

palladium 13, 15, 36, 55, 72, 98, 101–104, 124, 125
paramagnetism of small metallic particles 72
Pariser–Parr–Pople method 114
partition functions 93
Pauli paramagnetism 72
pentagonal bipyramid 94, 95, 97
photoelectron spectroscopy (PES) 148, 153 ff, 175, 187
platinum 13, 15, 34, 36, 37, 41, 51, 54, 65, 66, 82, 85, 125, 127, 133, 136, 150, 171, 176, 179
polydisperse systems 15, 71
polyhedra of pentagonal symmetry 92, 94, 95
polynuclear metal complexes, see metallic molecular clusters
populations of particles 28
Porod formula 81
radius 82
potassium 49
potential energy of clusters 91 ff
Lennard–Jones 91
Mie 91
Morse 91
Polanyi 169
powder metallurgy 13, 31, 38, 39, 62, 63
promoters, structural 34

random variable 18, 25
Raney (skeletal) nickel 35 ff
Rayleigh scattering 74, 75
coefficient 75
reduction 13, 31, 33, 34, 53
equilibrium constants of 32
reconstruction of metal surfaces 150
refractive index 74
repulsion integrals 108
residence time 46
resistance (resistivity) electrical 61–53, 174, 175
resolving power of microscopes 14, 157
rhenium 15, 36
rhodium 15, 34, 55, 124, 171, 190
Richardson–Dushman equation 64
ruthenium 13, 15, 34, 36, 55
Sabatier rule 183
saturation magnetization 67, 68

scattering of light 73, 74
Scherrer method 77, 79, 80
Schroedinger equation 107, 116
selectivity, catalytic 182
self consistent field scatered wave method (SCF-X_α-SW) 103, 115, 180
self consistent Hartree–Fock–Slater (SCHFS) 104 method 104
self consistent charge and configuıation (SCCC) method 113
semiconductors 126, 127
sessile drop method 121
shape factor of a particle 17
silver 13, 15, 36, 41, 48, 62, 66, 97, 100–102, 122—124, 127, 179
Raney 35
sintering 39
skewness of the distribution 20
Slater orbitals 117
small angle X-ray scattering (SAXS) 80 ff
Smoluchowski model of aggregation 127 ff
sols of metals 12, 36, 37, 41
sphericity of a particle 17
sputtering 41, 47
stability conditions for clusters 88 ff
stacking faults 79
standard deviation 19
sticking coefficient 46, 136, 139
stress in small particles 79
strontium 12
superparamagnetics 69
supersaturation 99
surface potential 162
segregation 157
susceptibility, magnetic 69 ff
synchrotron radiation 85

tantalum 164
thallium 12
thermal accomodation coefficient 139
thorium 34
tin 12, 39, 123
titanium 34, 123
Topping relation 165
transition state (activated complex) 183
treshold parameter, see location parameter
tungsten 13, 15, 155, 164, 177, 178
tunneling, quantum mechanical 64, 66
turnover number 182
twinning of tetrahedra 95, 97, 98
Tyndall effect, see Rayleigh scattering

ultramicroscope 74
ultraviolett photoelectron spectroscopy (UPS) 104
undemanding reactions — see facile reactions
uranium 134

van Arkel-de Boer process 39
vapour pressure of metals 42 ff
vibrating capacitor 163

work function 64, 100, 148, 152, 155, 161 ff

$X\alpha$-potential 116

Young equation 120

zeolites 53 ff
zero differential overlap (ZDO) approximation 110
zinc 12, 41, 122, 123, 186
zirconium 39, 123